Madness and Memory

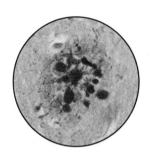

MADNESS
AND
MEMORY

The Discovery of Prions—
A New Biological Principle
of Disease

STANLEY B. PRUSINER, M.D.

Yale UNIVERSITY PRESS/NEW HAVEN & LONDON

Yale University Press books may be purchased in quantity for educational, business, or promotional use. For information, please e-mail sales.press@yale.edu (U.S. office) or sales@yaleup.co.uk (U.K. office).

Designed by Mary Valencia.
Set in Minion and Futura type by Westchester Book Group, Danbury, CT.
Printed in the United States of America.

Library of Congress Cataloging-in-Publication Data

Prusiner, Stanley B., 1942–, author.
Madness and memory : the discovery of prions—a new biological principle of
    disease / Stanley B. Prusiner.
        p.   cm.
    Includes bibliographical references and index.
    ISBN 978-0-300-19114-1 (cloth : alk. paper)
    I. Title.
    [DNLM: 1. Prusiner, Stanley B., 1942–.   2. Physicians—Autobiography.
3. History, 20th Century.   4. History, 21st Century.   5. Prions—history. WZ 100]
    R690
    610.92—dc23
    [B]

                                                                    2013041191

A catalogue record for this book is available from the British Library.

This paper meets the requirements of ANSI/NISO Z39.48-1992 (Permanence of Paper).

10 9 8 7 6 5 4 3 2 1

To scientific investigation that
creates new knowledge,
erases ignorance,
eradicates prejudice,
prevents disease,
alleviates suffering,
and enhances well-being

To Sunnie, Helen, Leah, and William

In striving to do scientific work, the chance—even for very gifted persons—to achieve something of real value is very small.
—Albert Einstein

Of all the services that can be rendered to science, the introduction of new ideas is the greatest.
—J. J. Thomson

The true men of action in our time, those who transform the world, are not the politicians and statesmen, but the scientists. Unfortunately, poetry cannot celebrate them, because their deeds are concerned with things, not persons, and are, therefore, speechless. When I find myself in the company of scientists, I feel like a shabby curate who has strayed by mistake into a drawing room full of dukes.
—W. H. Auden

The only good histories are those that have been written by the persons themselves who commanded in the affairs whereof they write.
—Michel de Montaigne

A man convinced against his will, is of the same opinion still.
—Samuel Butler

Contents

Preface

"What if I'm right after all?" I said to myself as I exited an elevator on the seventh floor of the Health Sciences East building at the University of California, San Francisco, one morning in October 1986. Four years had passed since my introduction of the word "prion" had elicited scorn, outrage, and disbelief from the scientific community. But one nugget of data after another was supporting my hypothesis—I was beginning to convince scholars that my novel perspective might have merit.

Thirty-five years after Thomas Kuhn wrote his remarkable analysis on the structure of scientific revolutions, Clayton Christensen wrote about the technological equivalent, which he labeled "disruptive technologies or innovations."[1] Christensen divided new technologies into two categories: sustaining and disruptive. Sustaining technologies rely on incremental improvements or small advances to nourish an established technology. In contrast, disruptive technologies are transformational and discontinuous. They create disorder by upsetting the status quo and are often greeted with profound skepticism. Some have destroyed entire industries—an example is digital photography, which Steven Sasson invented in 1975, while working at Eastman Kodak. On November 17, 2009, President Barack Obama awarded Sasson the National Medal of Technology and Innovation at a ceremony in the East Room of the White House. I had the privilege of meeting Sasson because I received the National Medal of Science on the same occasion.

Thirty-seven years after Sasson's disruptive innovation, Kodak filed for bankruptcy—its executives had numerous golden opportunities to capitalize on Sasson's invention; they possessed neither the necessary wisdom nor the uncommon vision. Digital photography did not fit the Kodak business model; how could the company survive if it were to abandon photographic film? Chemicals, film, photographic paper, and inexpensive cameras had become the heart and soul of a worldwide business with revenues of billions of dollars annually.[2] There is a strange irony to the Kodak story; from its founding by George Eastman in 1889, Kodak was a "high tech" business, yet it went broke under the burden of technological innovation.

In this book, I describe disruptive innovation in thinking about proteins and how they cause disease. I wrote this book because I feared that neither science historians nor journalists could construct an accurate narrative of my investigations. This is a first-person account of the thinking, the experiments, and the surrounding events that led to the identification of infectious proteins, or "prions" as I named them.[3] I have tried to describe what appears in retrospect to be an audacious plan to define the composition of the agent that causes scrapie, a barnyard disease whose etiology was a mystery at the time. On many occasions, I worried that my data might lead me down dead-end paths. Despite my fascination with the problem, I was haunted by a fear of failure; my anxiety was palpable at almost every turn. Was the problem intractable? As small successes emerged, so did a legion of naysayers who questioned both the wisdom of my pursuit and my scientific prowess; indeed, there were times when little but my naïveté and exuberance sustained me.

The skeptical and frequently hostile reactions to prions from many precincts of the scientific community reflected resistance to a profound change in thinking. Prions were seen as an anomaly: They reproduce and infect but contain no genetic material—neither DNA nor RNA; thus they constitute a disruptive transition in our understanding of the biological world. The consequences of the prion discovery are immense, and they continue to expand.[4] Their causative role in Alzheimer's and Parkinson's diseases has important implications for the diagnosis as well as the treatment of these common, invariably fatal maladies.

Readers may wonder about the title *Madness and Memory*. Prions clog the brain, causing neurodegeneration that often results in the diminished cognitive and intellectual function known as dementia. The word "dementia" is from the Latin *demens,* originally meaning "out of one's mind" or "madness." The etymology of *demens* is a derivation from *de-,* "without," plus *ment,* the root of *mens,* "mind." One of the more widely reported prion disorders was dubbed "mad cow" disease in the popular press. The most common form of dementia is Alzheimer's disease, which generally presents with memory deficits for recent events; such difficulties intensify as cognitive and intellectual failings worsen. Hence, *Madness and Mem-*

ory. The subtitle of this book, *The Discovery of Prions—A New Biological Principle of Disease*, comes from the 1997 Nobel Prize citation, but with one change that initially occurred to me during a joyous telephone call from Stockholm. This rather challenging moment is described in Chapter 16.

I am most grateful to my loving partner, Sunnie Evers, for encouraging me as I struggled to piece together a complicated story. For seven years, she has nurtured the maturation of this book. A sabbatical at Imperial College in London during the 2007–8 academic year provided the time to begin writing. Professor Chris Kennard was an extraordinary host, helping secure support from the Leverhulme Trust to defray the cost of my lodging and the editing of this book by Sara Lippincott, who converted my stilted prose into a readable story, for which I am very appreciative.

If I have erred in my recollection of the circumstances or in my analysis of others' conduct, I apologize. I hope that my attempts to describe the foibles of human behavior as they affected scientific investigation will not be misinterpreted as bitterness. I could have written a bland account of the events leading to the discovery of prions, but that would not have captured the passion with which my colleagues and I pursued our work. Nor would it have conveyed the complex emotions that permeate laboratories as scientists jockey for priority in the perennial race to be the first to discover the unexpected.

Many people have contributed to the studies described here. Their contributions are gratefully celebrated in the Acknowledgments. Any omissions are unintentional, and of course, I take full responsibility for any mistakes in the text.

Introduction

During a trip to Buenos Aires in September 1997 to participate in the Sixteenth World Congress of Neurology, I ventured into the city one afternoon. After buying a few presents for my two daughters, I wandered into an antique store on a side street. In this small shop full of old military uniforms and decorations, I spied a brass pocketknife. On one face of the handle were raised images of hand tools—pliers, a wrench, a drill, a trowel, shears, calipers, a hammer—and on the other engraved "ABOT herramientas," presumably the name of the tool company that produced it. This knife was a gem and I bought it for myself on the spot.

With my new pocketknife sequestered in my luggage, I left Buenos Aires for a meeting on the "Basic Mechanism of Prion Disease" in Stockholm. The day after arriving, I spent a few hours with Olle Lindberg, who had come to Stockholm by train from his farm in Dalarna. Olle and I had become good friends thirty years earlier when, still in my fourth year at the University of Pennsylvania's medical school, I worked in his laboratory at the Wenner-Gren Institute, then at Norrtullsgatan 16, in a drab yellow brick building with large wooden sash windows. It was during my time with him that I began to seriously consider the possibility of becoming a physician-scientist. Olle's warm and concerned counsel, his scientific wisdom, his gentle demeanor in the management of even the most difficult personalities, and his unwavering personal and professional ethics impressed me. Olle retired as director of the Institute in 1980, and soon thereafter he and his wife, Britta, moved to the farm where she had spent her childhood. Except during the harsh, dark winter months, Olle spent much of his time clearing underbrush from the thickly forested land, a task that gave him great pleasure.

As we reminisced and discussed my current research in a spacious corner room at the Diplomat Hotel with a view of the harbor, it occurred to me to show him the little pocketknife I had acquired in Buenos Aires. I was sure he would appreciate my purchase since he always carried one himself. Removing the knife from my toiletry kit, I proudly showed it to Olle, who admired it with a sparkle in his eyes and began thanking me. Though taken aback, I finally managed a few words, "Olle, I am so sorry! I didn't mean to give you the impression that it was a gift. I bought this in Buenos Aires for myself—it's a memento of the trip." As these awkward moments passed, Olle responded with his usual grace, but I was mortified. Each time I recall that episode, I marvel at how my brain continues to evoke the same feelings of surprise, confusion, embarrassment, and remorse.

Despite advances in unraveling the molecular basis of nervous system activity, our knowledge of the brain is still rudimentary. We have little understanding of the molecular events that occur when we experience events like the one I have just described—nor of what happens later when we remember the details with clarity. How did this little knife contretemps arouse such powerful feelings and stimulate my brain for years afterward with an intense cascade of recollected emotions?

Looking back, I think my fascination with the human nervous system began during a rotation on the neurology service as a third-year medical student at Penn. My teachers were distinguished and unforgettable: The chairman of the department was Milton Shy, who discovered what for many years was called Shy-Drager syndrome and is now referred to as multiple system atrophy (or MSA).[1] I would argue that the Shy-Drager eponym is a more evocative term than the nonspecific string of words "multiple system atrophy."[2] Ironically, Shy-Drager is a prion disease. Another remarkable teacher was a visiting professor, E. A. Carmichael, the former head of the National Hospital for Neurology and Neurosurgery at Queen's Square in London and a master clinician.

My next experience with neurology remains somewhat of a blur. During my medical internship at the University of California, San Francisco (UCSF), in the late 1960s, I spent a month on the neurology service, and while I remember the chief resident, I cannot recall the attending professor. I attribute this amnesia to the every-other-night call schedule. I was often so exhausted that it was nearly impossible to enjoy the flower-child culture of nearby Haight-Ashbury. I was envious of the relaxed lifestyle the hippies brought to the Bay Area, and I dreamed of having a house in Big Sur on a cliff overlooking the Pacific Ocean. I saw myself as an artisan of some sort, spending each late afternoon watching the sunset while reclining in a large, cast-iron, antique bathtub perched on the lawn. These restless thoughts were not new. A year earlier, during my stay in Stockholm with Olle, I had visited Morocco and fallen in love with the wonderfully preserved small town of Asilah, on the Atlantic coast about twenty miles southwest of Tangiers. Many of its stone ramparts and gateworks date back to the 1500s, when it was occupied by the Portuguese. After spending two nights in a magnificent old stone house at the edge of the Kasbah, I was ready to spend the next several years there, the romantic sound of waves breaking on the rocks as I drifted off to sleep only enhancing these impractical dreams.

But sanity, guilt, and lucid thinking prevailed, and I returned to Stockholm, where I continued my studies with Olle on the oxidative metabolism of brown fat cells. I was trying to figure out how these specialized fat cells could rapidly burn fat and generate lots of heat; normally, our bodies try to conserve energy and generate as little wasted energy in the form of heat as possible. After another two months of captivating research, I went back to Philadelphia to complete medical school, but I had become bedazzled by the "science bug." I was astonished that people actually got paid to solve puzzles every day—what a fantastic way to make a living! Although I wanted to return to Stockholm, this was not an option. The Vietnam War was raging in Southeast Asia, and I had two choices: become a U.S. Army hospital administrator or "suck-it-up" and do an internship before going to the National Institutes of Health (NIH). I briefly toyed with a third possibility—leaving the United States permanently—but in the end I was not prepared to constrict my future choices in who knows how many unforeseen ways.

While the grueling schedule of my UCSF internship is now a footnote, it left me with little energy or enthusiasm for clinical medicine. Fortunately, the next three years at the NIH working on enzymes in bacteria rekindled my interest in the study of human disease. One of my NIH colleagues, Mike Brown, would periodically ask me, "What are you going to do next, Stan? You're a physician; you need to figure out what you're going to do with your life. You've learned enough about enzyme regulation in bacteria and you're wasting time here." His advice was golden.

Once I decided to study the brain, I began to feel less adrift. With its billions of neurons, its ability to direct all aspects of human activity and its endless mysteries, the brain seemed a perfect subject for research, but

I had to pick a doable problem. I needed a focus with achievable goals for the next several decades—the more I thought about the brain, the more worthwhile my pursuit seemed. Before setting up a laboratory to study some aspect of the brain, I had to learn about the nervous system and define a research problem. As it happened, studies of the nervous system had reached a propitious point; the Society for Neuroscience had just been founded and was to grow over the next four decades to have more than forty thousand members worldwide.

After a few visits to laboratories studying the chemistry and physiology of the brain, I decided to pursue an abbreviated residency in neurology to gain the training I needed to become an effective physician-scientist. It was during my residency at UCSF that I encountered a patient with a rare, progressively debilitating illness called Creutzfeldt-Jakob disease (or CJD), and the mysteries surrounding this illness launched my scientific studies for the next four decades.

Neurologists have a vibrant, colorful, and almost endless vocabulary to describe normal functions and pathological states of the human nervous system. They are fond of words beginning with the negative prefix "a" from the Greek, such as "akinetic" to indicate that some motor functions that have become impaired; other such "a" terms include "anomia," "anosmia," "agnosia," "anosognosia," "aphasia," "apraxia," and "ataxia," and the parade of these wondrous words marches on. And then there is the "dys" prefix, meaning bad or abnormal; such words include "dysaesthesia," "dysautonomia," "dysphasia," "dysphemia," "dysmetria," "dysphonia," and "dystonia." Some neurologists seem to enjoy the cadence and rhythm that such terms as "perseveration," "dysdiadokokinesia," "dentatorubral-pallidoluysian atrophy," and "palatopharyngolaryngo-oculodiaphragmatic myoclonus" summon forth.

I was unprepared for the resistance of the scientific community to the discovery of prions—this created a rather harrowing and arduous journey for more than a decade. In the chapters that follow, I describe my thinking and that of others in elucidating new principles of disease that underlie many if not all neurodegenerative illnesses. Besides CJD, these disorders include Alzheimer's, Parkinson's, Lou Gehrig's (ALS), and Huntington's diseases, as well as the frontotemporal dementias (FTDs), including posttraumatic forms called dementia pugilistica and chronic traumatic encephalopathy (CTE). This book is a first-person account written by a scientist who had a leading role in these studies. As such, it must be considered a retelling of a scientific odyssey from my perspective that took me from pariah to prophet.

The current state of prion research, which continues to evolve and expand with unexpected discoveries, might best be described by quoting Winston Churchill. In late 1942, Churchill spoke at a Lord Mayor's luncheon in London, where he described recent British military success in North Africa but cautioned, "Now this is not the end. It is not even the beginning of the end. But it is, perhaps, the end of the beginning."

As one after another mysterious property of the prion began revealing itself, I was forced to think in ways that were often counterintuitive. Such reasoning is generally wrong, but occasionally it heralds a groundbreaking discovery of the sort epitomized by prions. My ideas contradicted the scientific body of knowledge: All biological entities, from viruses to people, have DNA or RNA, the genetic material that directs the synthesis of their progeny. Many argued that I was spewing heresy and I had to be wrong. The incredulity of my colleagues only strengthened my conviction that scientists have a responsibility to convince their skeptics of the validity and importance of discoveries that run counter to prevailing opinions, and that they can do so only by performing experiments that challenge their own hypotheses.[3] Sometimes the road of testing and retesting is long and arduous—such was the case for me.

Any commentary on scientific discoveries must include a discussion of luck. Extremely intelligent men and women can toil for years in the vineyards of science and never be fortunate enough to make a great discovery. And then there are a few people (I include myself) who are the recipients of a mammoth dose of good luck. The infectious pathogen that we now call a prion might well have turned out to be an atypical virus— not nearly as interesting as an infectious protein. Alternatively, prions might well have been much rarer than they are, and all but impossible to isolate; had that been so, we might have been unable to identify the protein of the prion. Or another group instead of mine might have discovered prions; that sort of preemption happens all the time in science—and unlike sporting events, there are no rematches. A discovery is made only once; there are no silver medals, no runner-up awards. Another championship game may be contested in twelve months, another Olympics in four years, but the race to a particular discovery is run only once.

Thirty years after the discovery of prions, we have come to appreciate their wide implications and the new challenges they pose. All neurodegenerative diseases are fatal; moreover, diagnostic tests are poor and effective therapeutics nonexistent for all of them. The discovery of prions has given us an entirely new approach to the development of drugs for treating these devastating maladies.

So how exactly did the story begin?

Author's Note

The book that you are reading contains a true story. The conversations in the book all come from my clear recollection of them, though they are not written to represent word-for-word transcripts. Rather, I have retold them in a way that evokes the feeling and meaning of what was said. In all instances, the essence of the dialogue is accurate.

Madness and Memory

1

Growing Up

That I chose an academic career seems a bit odd, for I found both grade school and high school rather boring. I could get Bs with little effort, and those always seemed good enough. My parents never pushed me to do better, even though both of them had been A students.

My junior and senior high school was Walnut Hills High School in Cincinnati, Ohio. It was a college-preparatory public high school; you had to take a test in the sixth grade to gain admission. But Walnut Hills, as far as I could tell, was no better than elementary school: many of its teachers had been there for decades and seemed to have little interest in capturing the attention of their students. Since we were required to take Latin for three years beginning in the seventh grade, my father decided to tutor me so that I could more easily grasp this arcane language. His tutelage helped, and Latin became one of my favorite subjects—once I had mastered the fundamentals. In fact, it was so easy that I took Latin for five years instead of adding a more useful language like French, a decision I'd later regret.

My high school experience with the sciences was somewhat different. I took advanced-placement math courses my senior year, but my homeroom teacher, who taught chemistry, would not allow me to take AP chemistry. When my parents complained, he told them that I would never be able to comprehend the science. Instead, I took the non-AP chemistry course and wound up tutoring many of my friends, who found the abstract thinking in chemistry difficult. Did I major in chemistry at the University of Pennsylvania just to prove my homeroom teacher wrong? Did I spend my life doing biochemical research in quest of his praise? A few psychiatrists might choose that interpretation, but I doubt that this high school rebuff had anything to do with my pursuit of chemistry. I enjoyed chemistry because I never had to memorize anything. The key was to balance the equations; I had to be sure that each side of the equation had the same number of hydrogen, carbon, nitrogen, oxygen, and sulfur atoms. Once my addition was correct, all I needed was some rudimentary knowledge of the process—though I have since learned that chemical research demands a bit more savvy and a lot more insight.

In spite of my uninspiring classes, I was enthusiastic about the Boy Scouts and my high school fraternity. I preferred scouting activities to those demanded of me as a student, and they provided an antidote for my boredom. I became an Eagle Scout and took summer jobs at the local scout camp. I even tried to postpone college by applying for a yearlong position as a Boy Scout–military aide at the U.S. Air Force base at Thule, Greenland. Only many years later, after a visit to the South Pole, did I realize how lucky I was to have been turned down.

Despite having never made it to Greenland, I retained an interest in cold temperatures. My first research experience began during the summer after my junior year at Penn, when I investigated the effects of hypothermia in retarding brain swelling in rats.[1] The next summer, I pursued a project directed by Bernard Black-Schaffer at the University of Cincinnati, in which I studied the tolerance of hypothermic hamsters to acceleration forces of more than eight hundred times that of gravity, such as might be experienced during interstellar space travel.[2] During medical school at Penn, I began studying the brown fat of hamsters during arousal from hibernation.[3]

While my infatuation with cold temperatures was to end with my brown fat studies, my love of science grew. The privilege of spending time discovering something that no one else had ever known before became an insatiable thirst. As it happened, hamsters would continue to play a critical role in my research as the mysteries of the prion unfolded.

My Forebears . . .

The roots of my love for science have always been mysterious to me. Neither of my parents was a scientist, but they must have bequeathed to me certain traits useful in that field. My father, Lawrence Albert Prusiner, was an architect with an aptitude for numbers, and my mother, Miriam Hannah Spigel, was imaginative, creative, and relentless in pursuing her passions (pottery, tropical fish, bonsai, ikebana). Both were descendants of Jewish immigrants from eastern Europe. To understand my parents better, I became obsessed with learning about our origins (figures 1–2).

My paternal great-grandfather Wulf Prusner (the i crept in later) was a prominent attorney in Moscow, which he left in the spring of 1891, shortly after its governor, the brother of Czar Alexander III, ordered the expulsion of Jews from the city. Permission to remain was given only to those who converted to Christianity, to those Jewish women willing to become prostitutes, and to those wealthy Jewish merchants and their families deemed useful to the city's economy. Altogether, more than fourteen thousand Jewish families were expelled from Moscow to the Pale of Settlement, a region of western Russia that included much of present-day Lithuania, Belarus, Poland, and Ukraine.

In Moscow, Wulf managed to obtain a passport from the German consul general. While his wife, Feine Moiseev, and their two sons and one daughter moved to Mogilev, the capital of eastern Byelorussia (Belarus), he traveled to Germany, intent on emigrating to the United States to make a better life for his family. He arrived in New York and made his way to Chicago, where he got a job rolling cigars. Learning that better jobs were available farther west, he moved on and by the end of 1891 he had reached Sioux City, Iowa, where there was a growing, vibrant Jewish community. The earliest Jewish settlers in Sioux City predate the Civil War. All were of German descent; the first Russian Jews arrived there in 1888, only three years before Wulf. There, he took a number of menial jobs: night watchman at a streetcar barn, janitor at a public library, fruit peddler. He went to night school to learn English; eventually he was to build an insurance business, which, having Anglicized his first name, he called the William L. Prusiner Insurance Company.

As an attorney in Moscow, Wulf had managed to amass the equivalent of more than $40,000 in assets, a small fortune in those times. When the Jews were expelled, he gave power of attorney to a trusted friend, who agreed to liquidate his properties and give the proceeds to Feine and their three children. A few months after her banishment to Mogilev, Feine and the children went back to Moscow to collect her husband's estate from his

Figure 1   Prusiner side of my family. Clockwise from upper left: my great-great grandfather Lippman-David Prusiner, father of Wulf; Feine (seated left) and Wulf (center), my grandfather Ben (standing left), Dave (standing right), Molly (seated right), and Bessie (bottom); my grandmother Ethel Galinsky with her sons, my father Lawrence (right) and his brother Stanley; me (right), my father (center), and brother Paul in 1997; and my great grandfather Moshe Lazar Galinsky, father of Ethel.

Figure 2 Spigel side of my family. Clockwise from upper left: my grandfather Ben (standing) and grandmother Mollie Feldman (seated) with their children, Herbert (left), Miriam (center), and Naomi; my father and mother; me at age three; my mother and father at ages fifty-two and fifty-six.

friend, who treacherously notified the police that they had returned to the city. A regiment of officials swooped down on them and thrust them into jail. For making a faint stand for her rights, Feine was fined the equivalent of $500. In a letter to her husband in Sioux City, she wrote that this was "all the money I had on earth." After paying the fine, Feine and her children were taken to the outskirts of Moscow and ordered to leave the country. Wulf sent his wife $125 for passage to America, and the family started for the German frontier, four days' journey by rail. When they reached the border, they were ordered back by sentries, who were there to stem the tide of Russian-Jewish refugees. It was a time of mass exodus: Every day, nearly four thousand Russian Jews tried to leave. In despair, Feine made her way back to Mogilev to wait.

According to a contemporary article in the *Sioux City Journal,* when two of Wulf's Sioux City friends lobbied for transit visas for Wulf's wife and children, they were told that "on account of the German regulations and quarantine laws, . . . it [would be] impossible to accomplish anything." But they appealed to the U.S. State Department on Feine's behalf. Their effort was rewarded in March 1893, when Secretary of State Walter Gresham received a dispatch from the American legation in Berlin: The German government had granted Feine Prusiner and her children permission to pass through Germany. By mid-April, the family had left Hamburg on the ship *California,* arriving in New York at the end of the month. Shortly afterward, they were reunited with Wulf in Sioux City.

Four years later, Wulf would bring his two sisters and his father, Lippman-David, to Sioux City. Lippman-David's father, Noach, is thought to have moved to Shklov on the Dnieper River from the town of Pruzhany, or Prusana, in western Byelorussia near the Lithuanian border. Presumably Noach had no surname, as was common for Jews living in the Pale. Toward the middle of the nineteenth century, that changed when the czar decreed that Jews had to take last names and that those living outside their birthplace could take the name of the town from which they came. It was probably Noach who chose the name Prusner, which has the Yiddish ending "ner"; Russian officials would probably have given him a "sky" or "ski" instead.

Benjamin, the eldest child of Wulf and Feine Prusiner, was my grandfather. He never finished high school; he was forced to work to help his family survive. Eventually he built a small firm, the Benjamin W. Prusiner Insurance Company, in Des Moines. Benjamin's wife, my paternal grandmother, Ethel Galinsky, was born in Omaha, Nebraska, in 1884 and died at age thirty-two of a pulmonary embolism, four years after giving birth to my father and two years after delivering his brother, Stanley Arnold.

On my mother's side, my grandfather was also named Benjamin. His father, Moses Herman Spigel, and mother, Sarah Weinstein, were both born in Austria but moved to Romania, where they farmed tobacco before being encouraged by the State of Virginia to immigrate. In the Richmond area, Moses farmed tobacco for many years and brought his father, Benjamin, and mother, Fannie Betseig, from Austria. Moses and Sarah had eleven children; my grandfather Benjamin Spigel was the oldest of nine boys and was born on Christmas 1878, on shipboard as his parents and his sister Hannah crossed the Atlantic. After raising their family in Richmond, Moses and Sarah moved to Norfolk, where their son Benjamin had built a successful jewelry business and become a member of the city council. I never met my grandfather Benjamin Spigel because he was killed in a streetcar accident at age fifty; at that time, my mother was only fourteen years old. She had adored her father, and her memories of him remained vivid and indelible for six decades. On the day of his death, the family's rabbi wrote:

> The loss of Benjamin Spigel . . . produced a void that cannot be filled. . . . To him came the orphan and the widow, the sorrowful and the oppressed, the ill and the troubled, for aid in their perplexing problems. To him came Jew and Gentile, big and little, for the sort of help that comes not only from the purse but also from the heart. . . . Benjamin Spigel was not only Norfolk's finest Jew and citizen; he was one of God's noblemen. . . . He was an ambassador of hope and an apostle of human love.

Benjamin Spigel's wife, Mollie, was my beloved maternal grandmother; she was born in 1887 in Salisbury, North Carolina, the second of six children of Jacob and Leah Feldman, both of whom came from Vilnius, Lithuania, at the end of the nineteenth century. When Jacob came to New York, he encountered the same problem as Wulf Prusiner had: too many immigrant Jews and not enough jobs. Jacob went south, to Baltimore, to stay with a cousin, but no jobs would be found there either. Encouraged to go farther south, he bought a horse and a wagon on whose sides were painted the words "Feldman and Sons," so Jacob took the name "Feldman" from his wagon. None of his descendants know what his surname was in Lithuania or the name he used when he entered the United States. Like many eastern European Jews, Jacob and Leah wanted to forget a life that was harsh and filled with state-sponsored anti-Semitic violence. Jacob built a thriving dry goods business, with six stores in North Carolina and southern Virginia, but lost nearly everything in the Depression. My mother grew up in Norfolk; after her father's untimely death, there was enough money to send her to Hollins College, but only for two years.

My father, on the other hand, graduated from Iowa State University and received a master's degree in architecture from Harvard. Though jobs were scarce in the Depression, he found work as an architect in Washington, D.C., where he met my mother, who was working as a secretary. They postponed their marriage for more than a year when my father's brother developed Hodgkin's disease. His stepmother demanded that my father return to Des Moines to care for his dying brother. Eventually, he and my mother married on February 18, 1939, and settled in Des Moines, where I was born on May 28, 1942. I was named for my dead uncle Stanley Arnold but rather considerately given the middle name Ben so my initials would not be S.A.P.

Shortly afterward, my parents moved to Boston so that my father, who had enlisted when the Second World War broke out, could attend Naval Officer Candidate School before being sent to the island of Eniwetok, in the South Pacific. At that point, my mother took me to Cincinnati and moved into the same ten-story, red brick Presidential apartment building on Greenwood Avenue as her mother, Mollie Spigel. The hallways were filled with the stale odors of cooked food from kitchens that had little or no ventilation. Our small two-room apartments—my mother and I lived on the seventh floor and my grandmother on the fifth—had pull-down Murphy beds that were stored by day in shallow closets. The proximity of my grandmother's apartment was my great fortune. Whenever my mother became cross with me, which was rare, I needed only to go down two floors to see my loving grandmother. In her eyes, I could do no wrong! Moreover, for several years I didn't have to share either of these two women, both of whom doted on me. Their unconditional love created a level of self-confidence that was probably responsible for the work I was able to accomplish in later life. My father was overseas for two years, allowing me to be the little man of the house.

I still hold many loving memories of Granny. Her homemade chicken soup was fantastic—a recipe passed on to my mother. She and my mother read to me every day for hours; there were no television sets. Remembering my love for these two women who gave so much of themselves to me can still bring tears to my eyes. And the deep sadness I felt at age twelve when Granny died at sixty-seven of a hypertensive cerebral hemorrhage is among the most vivid of my childhood recollections.

During the war, my grandfather Ben Prusiner would sometimes visit us on Saturday mornings for brunch; he lived seventy-five miles away in Springfield, Ohio, where he managed a dry goods business and struggled to maintain his insurance business in Des Moines. My mother would prepare a sumptuous meal, of which the constant was pickled herring immersed in sour cream topped with chopped onions and capers. Whether the herring was layered on toasted bagels or slices of rye bread is a detail that escapes me; perhaps my love of toasted poppy seed bagels has its origin in those brunches with my grandfather. Ben died at the age of sixty-five, after his second heart attack.

Shortly after my father's return from the South Pacific, we moved back to Des Moines. Two months after my brother Paul's birth in 1948, my mother was hospitalized for many weeks. Her absence created an unfillable void in my world; I had just turned six. Two years later, she suffered her first heart attack, at age thirty-four. She was hospitalized for several months, during which I was allowed to see her only once and Paul, at age two, was not permitted to enter the hospital at all. I missed her terribly—she was the center of my world, her voice and smile bringing me so much joy and warmth. My father's love couldn't fill the emptiness created by her absence. My father visited her every day and had to care for us as well—which may have contributed to his relative lack of success as an architect. He was a selfless man who cared deeply about his family.

For the next forty years, I was to remain concerned about my mother's health. Her illness was never far from the minds of her husband and two sons. I have no doubt that the tenuousness of my mother's existence shaded our views of life. This uncertainty intensified the problems emanating from my father's inability to find a high-paying job as an architect. He and my mother struggled each month to pay their bills; rarely were they able to save any money. Both of my parents remained frustrated by their modest resources, but neither knew how to break out of the situation. They felt trapped.

We always lived in decent houses, but never in the best neighborhoods. Our first house was a small, two-bedroom, newly built house on Hillside Avenue in West Des Moines, just like several dozen other new houses in a tract near the city's eastern border. It was modest but comfortable, with enough land for a sizable garden in the backyard. Each evening after dinner, I would perch on a dining room chair next to the buffet and snuggle up to a small brown radio tuned to *The Lone Ranger*. My world changed dramatically when my father's stepbrother sent us our first television. Its six-inch screen was just fine for watching the Lone Ranger galloping along on his white horse, Silver, with his Indian sidekick Tonto close behind. We lived there for several years before moving to a larger house on the western edge of Des Moines proper, after my mother's heart attack.

Although now I had my own bedroom, with a door that opened directly onto a big backyard, my memories of this house are mixed. My mother's illness cast a shadow on it. My parents, particularly my mother, seemed more sensitive than ever to their relative poverty and to the inadequacy of our home. My father was a rock, however, year after year plying his skills as an architect, designing schools, hospitals, and industrial buildings. His work was tedious and exacting; he was always exhausted when he came home. Curiously, I cannot recall ever seeing him bring work home. He seemed diligent about separating his work from his family life.

Most of my parents' friends were Jewish. The Jewish community in Des Moines consisted of fewer than one thousand families, and my parents were at the lower end of the socioeconomic scale in this small but prosperous community. Their plight was made all the more evident, to themselves and others, by the comparable wealth of close relatives, none of whom seemed to care that my father was struggling to hold his family together. Of the public schools I have little to say, beyond that I cannot today recall a single one of my teachers. In 1952 we moved back to Cincinnati, where, in quest of a better job, my father joined the firm of A. M. Kinney Architects and Engineers. He spent the remainder of his professional life there, until it collapsed during the 1974 recession, putting him out of work at age sixty-two.

A sadness pervades many of my reflections on my parents' lives. Both had difficult childhoods: My father lost his mother at age four, and my mother lost her father at age fourteen. My father never spoke about his mother, so I surmise that he had few memories. I can't recall him or my mother ever describing his stepmother in positive terms, except to praise the peanut brittle she made for the holidays. My mother's childhood was probably happier than my father's, but she had to contend with a difficult, dull older sister and a talented, aloof older brother. Both my mother and her sister tired of their mother's adoration for her son, Herbert, who trained as an architect at the University of Pennsylvania. Herbert left architecture for the steel scrap business and died of gastric cancer at the untimely age of thirty-six.

My parents' adult lives were no less arduous than their childhoods: They lived from one paycheck to another. They were unable to save any money—their only nest egg was our modest home in Cincinnati, once the mortgage had been paid off. With my mother's chronic health issues compounding our penury, my father's selfless focus on keeping his family together seems all the more remarkable. When life became too challenging, he would retreat to his bedroom and read mystery novels.

As I try to understand how those years molded my personality, I contemplate a paradox: I occasionally struggled with low self-esteem, yet I never lost confidence in my ability to do anything I might choose. Ascribing the low self-esteem to my father's apparent lack of accomplishment, or to growing up Jewish in mostly Gentile Des Moines, seems too simplistic. My unanalyzed self-image remains mysterious to me.

My sense that I was odd, a bit different, lessened as my ties to Judaism strengthened when I moved eastward: first to Cincinnati and later, in 1960, to Philadelphia, to matriculate at the University of Pennsylvania. My widowed aunt, Helen Spigel, lived in Philadelphia and encouraged me to consider Penn and come live with her and my three cousins. Cincinnati and Philadelphia were both major centers of American Jewish culture and learning, and my early feelings of estrangement from American society slowly shifted to an embrace of my heritage. Although I liked Cincinnati much better than Des Moines, I had found high school dreary, and I yearned for something better. I found it at Penn; my classes—chemistry, biology, economics, history of architecture, philosophy, Russian history—were stimulating, and I enjoyed my professors and respected their research. I could sense their enthusiasm for their chosen fields, and in no time I was drawn by their example into the excitement of intellectual inquiry.

My Scientific Mentors

I was extremely fortunate to have almost a dozen superb mentors (figure 3). Beginning in the summer before my senior year at Penn, I studied brain swelling in rats with Sidney Wolfson, and the experience persuaded me to stay on at Penn and enter the medical school there. Wolfson showed me how to read scientific papers and analyze data. He taught me statistics, and how to formulate a scientific problem. He spent an immense amount of time helping me delve into the fascinating world of scientific research. After two years working with Wolfson, I shifted my research efforts to the oxidative metabolism of brown fat and began studies, with Britton Chance and C. P. Lee, focused on heat production in the brown fat of hamsters as they roused from hibernation. Chance was one of Penn's most renowned scientists—an authority on energy metabolism. He was one of those phenomenal scientists who virtually inhale new information, but he was impatient and always seemed to be in too much of a hurry. I managed to isolate brown fat cells and found I could stimulate their metabolism by adding the hormone noradrenaline, which resulted in the oxidation of the fat and the concomitant production of heat. C. P. Lee, a tall woman with a round face and a big laugh, tutored me in how to conduct scientific investigations. Her influence on me was immense. Her wonderful sense of what is important was reflected in her sage advice: "Stan, don't be upset about someone competing with you and repeating your work. That means your work has sufficient merit that someone would independently undertake to repeat your experiments in the context of their own studies. Always remember, no scientific result is considered valid until it has been independently confirmed."

I continued my research on brown fat cells with Olle Lindberg for much of my fourth year of medical school, from September 1967 to February 1968. Lindberg was the director of the Wenner-Gren Institute, which, like Penn, was a mecca for research on energy metabolism. Despite our vastly different stations in life, Lindberg treated me as an equal, and I grew fond of him. Along with discussing the latest results of our work, we would talk about science, politics, and philosophy—and the personnel problems that now and then rudely interrupted our scientific pursuits. One denizen of the institute so exasperated Lindberg that he shared a shrewd adage: "When a wise man speaks to a crazy man long enough, the

Figure 3   Some teachers and mentors. Clockwise from top left: Sidney Wolfson, Olle Lindberg, Earl Stadtman, Britton Chance, and Charles Yanofsky.

wise man is no longer sure who is crazy." (Neither Lindberg nor I have ever been able to establish the origin of this useful philosophical nugget.) My time with Lindberg convinced me that the life of a scientist could be an interesting and enjoyable way to earn a living. And I might even be reasonably good at it!

Back at Penn in the spring of 1968, I was faced with some difficult decisions. My medical school studies were coming to a close, and I yearned to go back to Sweden. I wanted to continue my studies of brown fat, sure that there was more to be learned and that I could make a substantial contribution. But in the fall of 1967, I had applied to the medical internship-matching program, and my next step was going to be an internship in medicine.

The medical internship at UCSF had become one of the most prized in America under Lloyd Hollingsworth (Holly) Smith, who had taken the chair of medicine only four years earlier. While I was fortunate to have been accepted, the internship proved every bit as grueling as I feared. After taking care of patients for thirty-six hours, I barely had time to gather my strength before reporting the next morning for another thirty-six-hour siege. (Some years later, the every-other-night training programs were eliminated, because neither patients nor interns benefited.) My minimal experience in caring for sick people was evident from the first moment. I had much to learn, and I was frightened that I might kill a patient. But I had no choice; I had to perform. I couldn't quit; otherwise, I would have been immediately drafted and shipped off to Vietnam as a field hospital administrator.

After a year of torment, I left San Francisco for Bethesda, Maryland, to report to the NIH, where I had been accepted as a research associate. Out of nine possible positions at the NIH, I had recklessly decided that only one was sufficiently interesting to apply for—such hubris! Even being an army doctor in Vietnam would be preferable, I thought, to enduring a two-year stint in some boring lab. Fortunately I was awarded the one position I had applied for, at the National Heart Institute. In July 1969 I reported to the NIH as a lieutenant commander in the U.S. Public Health Service and spent the next three years in the laboratory of Earl Stadtman, an outstanding biochemist specializing in enzymes.

My first day at the NIH was a bit rough. When I walked into Stadtman's office to say hello, he greeted me by saying, "Oh, I forgot you were coming," a remark that did little to boost my morale. An introvert, Stadtman found conversation laborious unless it was after five and some outstanding bottles of wine were being consumed; he was a genuine oenophile, with an encyclopedic knowledge of French wines. His shyness was difficult for many of the postdocs, because it was often interpreted as a lack of interest in you and your substandard work. While this was generally not the case, the insecurity of a bunch of high-achieving doctors trying to do science only served to reinforce the notion that our work was marginal.

As I recall those days now, I see that Earl was unbelievably patient with me and several other behaviorally challenged scientists in his care. There I initially studied an enzyme called glutamine synthetase, which catalyzes the synthesis of the amino acid glutamine; and later, glutaminase, which acts in the breakdown of glutamine. Despite my earlier research experiences, I soon realized that I knew rather little about scientific investigation. Stadtman taught me the fundamentals of research and the art of purifying proteins. Fortunately, his tutelage eventually turned out to be a primer course on the isolation and identification of the elusive prion.

2

The Beginning of an Odyssey

After three years at the NIH, it was time to move on. I felt ready to build my own laboratory, but I somehow managed to grow up and take what in retrospect was a remarkably mature approach to gaining the education that I would need for studying the brain. I chose to undertake an abbreviated residency in neurology before setting up a laboratory. Most amazing, I still didn't have a clue about what I wanted to study. My goal was to identify, during my residency, a great problem for investigation.

In early September 1972, I was performing a lumbar puncture (or spinal tap, as it is often called) when the hospital page operator summoned me: "*Dr. Prusiner, Dr. Stanley Prusiner...*" As soon as I finished the procedure, I called in and learned I had been assigned a patient on Dr. Donald Macrae's service, admitted for evaluation of a rapidly progressive dementia.

When I entered her room, my new patient, dressed in a print hospital gown, was sitting on the side of the bed. She was well-tanned, slim, with straight, graying brown hair cut just above the shoulders and deep blue eyes that glowed with enthusiasm. She seemed calm and somewhat detached. This beautiful sixty-year-old woman from Marin County, north of the city, was having difficulty with her memory and the fine movements of her fingers. She had problems finding words and was unable to describe the symptoms that had brought her to the hospital. Her husband recounted the initial difficulties that had alerted them, such as her inability to insert her car key into the ignition or to unzip the ball compartment on her golf bag. During my mental-status examination, I found that her ability to remember recent events was poor; according to her husband, these memory deficits were new. While I was testing her gait, muscle strength, sensory modalities, and reflexes, she exhibited myoclonus—jerky movements in her muscles. I had no idea what had caused her to worsen so quickly.

Progressive dementia—which involves a loss of intellect as well as memory—combined with myoclonus prompts consideration of Creutzfeldt-Jakob disease (CJD). When the senior resident proposed this diagnosis, I vaguely remembered an affliction I had read about in medical school called Jakob-Creutzfeldt disease, but this was my first year of residency, and I had had no clinical experience with it. (I later learned that the names were transposed, for the most idiotic of reasons, in the short interval between my medical school studies and my residency.)

CJD is invariably fatal. It is one of a cluster of human and animal diseases now called prion diseases, the most familiar of which is bovine spongiform encephalopathy (BSE), or "mad cow" disease. Another is scrapie, which affects sheep and goats (the name arose from the tendency of infected animals to scrape off their wool or hair in an attempt to relieve severe itching). Typically, the brains of CJD patients, like those of livestock afflicted with scrapie or BSE, contain numerous holes ("vacuoles"), which give the brain a spongy ("spongiform") appearance under the microscope (figure 4). Sometimes these illnesses are referred to as the transmissible spongiform encephalopathies, or TSEs.

The etiology of CJD—exactly what was transmitted from one patient to another—was unknown, but a "slow virus," one that causes disease only after many months or even years of incubation, was suspected. C. Joseph Gibbs, D. Carleton Gajdusek, and their colleagues at the NIH had shown the disease to be transmissible by injecting brain extract from a dead CJD patient into a chimpanzee, which developed symptoms of CJD thirteen months later.[1] These same researchers had also experimentally transmitted another TSE, known as kuru, a fatal neurological disease found in the Fore people of the Eastern Highlands of New Guinea, to chimpanzees.[2] Like CJD, kuru was thought to be caused by a slow virus.

To a busy first-year neurology resident, the viral explanation seemed adequate. But as I watched my patient deteriorate rapidly over the next

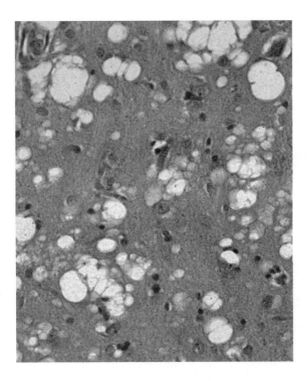

Figure 4   Brains of CJD patients, like those of livestock afflicted with scrapie or BSE, contain numerous holes ("vacuoles"), which give the brain a spongy ("spongiform") appearance in the microscope.

few weeks, I was perplexed. She exhibited no signs of an infectious disease: Her neck was supple, not stiff; she did not have a fever; and she had no increase in white blood cells in either her blood or her cerebrospinal fluid.

Besides being seen by Macrae, an outstanding clinical neurologist and celebrated teacher, my patient was also examined by Robert Fishman, chairman of UCSF's Neurology Department (figure 5). Both agreed that she had CJD and was likely to die within a few months. When Fishman, a superb clinician and teacher but with no particular expertise in the dementias, CJD, or viral diseases, rotated off the service, another senior neurologist, J. Richard Baringer, became the attending physician. Baringer was interested in viral infections of the brain; his research was focused on the persistence of herpes viruses in the nervous system, and slow-acting viruses were thought to cause herpes infections as well as

Figure 5 More teachers and mentors. From left: Robert Fishman, Ivan Diamond (right of unidentified woman), Donald Macrae.

CJD. Our conversations about CJD grew provocative, and he referred me to papers by the radiation biologist Tikvah Alper, who had described the strange properties of the "scrapie agent" infecting sheep and goats, which was thought to be similar to the slow virus causing CJD.

Alper, who trained in Berlin with the nuclear physicist Lise Meitner in the 1930s, was an outspoken, feisty woman who grew up in South Africa, where her Russian-Jewish parents had immigrated. She moved to London in 1951, her forthright opposition to apartheid probably prompting the South African government to facilitate her exit. There she became director of the Medical Research Council's Experimental Radiobiology Unit at the Hammersmith Hospital in 1962 and continued in that position until her retirement in 1974. It was during this period that she published her seminal work on the scrapie agent (figure 6).

From her studies on the extreme resistance of the agent to inactivation by ultraviolet light and X-rays, Alper reasoned that it could not be a virus, since the high doses of ultraviolet light and X-rays would have

Figure 6 Tikvah Alper used ultraviolet light of different wavelengths to analyze the scrapie agent.

killed any virus by damaging its nucleic acid genome.[3] I concluded that Alper was unlikely to be correct; after all, every known infectious pathogen, viruses included, had a nucleic acid core that directed the synthesis of its progeny. How else could a pathogen flourish? Either DNA or RNA *had* to be involved, since they are the genetic materials of life. I thought a novel virus was the most reasonable explanation for Alper's results. What else could the "scrapie agent" be? There *was* nothing else.

My patient's hospitalization lasted almost three weeks, so I was able to watch her neurological status decline daily. Eventually she could not walk even with assistance. She left the hospital for a nursing home but returned four weeks later, unable to sit up, feed herself, or speak—she had become a neurological vegetable. The beautiful woman who had entered the hospital less than two months earlier was now mute and all but immobile, with no understanding of who she was or what had happened to her. I felt profoundly sorry for her—and for her husband, who could only stand by and watch his wife being destroyed by a "slow virus." Baringer and I were helpless—there was not a single medicine that could modify her outcome.

Three weeks later she died. At autopsy, Baringer removed her brain and, after examining small pieces of it under the microscope, confirmed the diagnosis of CJD, although the characteristic spongiform change was minimal. He sent some of her frozen brain tissue to Joseph Gibbs at the NIH, where it was homogenized and injected into the brain of a monkey that, sure enough, developed the disease two years later.

Over the next eight months, I read everything I could find about scrapie and CJD. I reread Alper's work several times, but it didn't clarify the question of etiology. If anything, I became more confused—and simultaneously more enthralled. The problem was beginning to capture my imagination.

The agents causing CJD in people and scrapie in sheep and goats were mysterious. They resisted killing by boiling or exposure to chemicals such as alcohols and formaldehyde, the principal ingredient of formalin. Formalin is a common fixative, used to harden tissues before cutting sections for microscopic examination. It is also used to inactivate viruses during the preparation of vaccines, such as (for example) the Salk polio vaccine; it renders the poliovirus harmless but allows the vaccinated individual to produce antibodies to the virus. In the 1940s thousands of sheep in Scotland were vaccinated with formalin-treated louping-ill virus (LIV), which causes meningoencephalomyelitis in livestock—only to develop scrapie a couple of years later. The crude preparations of LIV had been produced from a sheep that also harbored the scrapie agent. Though formalin had inactivated the LIV, it had not killed all of the scrapie agent.

In contrast to viruses that could be readily identified in the electron microscope, the scrapie and CJD agents remained elusive. The medical literature was replete with unsubstantiated claims and counterclaims about electron-microscopic sightings. Some investigators argued, like Alper, that the scrapie agent was devoid of nucleic acid and therefore could not be a virus. Others thought that her findings were fallacious and that the scrapie agent's resistance to destruction by ultraviolet light and X-rays could be explained by some undiscovered viral attribute—that the scrapie agent was an unusual virus, but a virus nonetheless.

Whatever the explanation, the scrapie agent was unlike any other infectious pathogen. I was baffled, my curiosity thoroughly aroused. Whenever others argued that the mysterious nature of the agent made it too difficult to study, I found myself more and more attracted to the challenge. What causes scrapie? I could not let it go. By the end of my first year of residency, I was obsessed by the inexplicable. As I discussed with colleagues the evidence for and against the hypothesis that a slow virus causes CJD, I grew convinced that a biochemical research program devoted to isolating and characterizing the scrapie agent might be a way forward.

Unlike most of my colleagues, who thought my desire to study scrapie was misguided, the eminent Stanford geneticist Charles Yanofsky was enthusiastic (see figure 3). Charley and I had first met at a dinner he and his wife gave at their home in Palo Alto for some friends we had in common. A tall, handsome man with extraordinary intellect, he had a somewhat halting gait. His feet had been frostbitten when he was a U.S. infantryman at the Battle of the Bulge; the frostbite had undoubtedly saved his life, since it prevented him from participating in further combat. Charley understood my vision, felt my excitement, and thought I should get started. When we met again, in my last year of residency, he was disappointed that I hadn't yet begun experiments to determine what causes scrapie. "I'm still taking care of patients," I explained. "I won't have time to begin those studies until next summer, when I finish."

"You need to get going," he roared. "You're wasting time!"

It was easy for Charley to understand what I wanted to do. As the discoverer of colinearity, the direct relationship between the structure of genes and that of their encoded proteins, he was a legend in molecular biology. I, too, wanted to wake up in the morning excited by unknowns in my research. I wanted to go to bed at night energized by the thought of making a breakthrough that would improve people's lives. I wanted to experience again the excitement I had felt in medical school in my experiments with brown fat metabolism. Based on that work, I was sure I could construct another important set of investigations. Identification of the scrapie agent was a problem that just might be worth committing all my intellectual energy to. The clues from Tikvah Alper's work that the scrapie agent was different from all other infectious pathogens were tantalizing. Many scientists likened the scrapie agent to kryptonite.[4] I couldn't imagine a better problem to tackle. It had all the hallmarks of a new frontier in neurologic medicine—a truly great scientific puzzle that cried out for solution.

Moreover, deciphering the riddle of "What causes scrapie?" might well lead to an understanding of CJD. By now, the triad of scrapie, kuru, and CJD had been well established by transmission and neuropathological studies. Defining the nature of the infectious agents causing these diseases was clearly the next important step. Scrapie was no longer the exclusive province of veterinarians; it had been catapulted to the center of human medicine by the transmission of kuru and CJD to apes and monkeys.

Every scientist must choose a problem; it is probably the single most important decision of his or her professional life. Yet few seem to do so with much forethought. Most choose a problem related in some way to their training. The argument for this is that granting agencies will not support you unless you have already worked in the chosen area. And why *not* work for an expert before setting out on your own? At first glance, it seems a reasonable way to proceed. Apprenticeships have been around for centuries; they work well in training doctors, lawyers, engineers, and other professionals—but they may not be the best formula for scientists or other creative people engaged in forging new endeavors.

Before beginning my neurology residency, I had decided for many reasons not to do any research during that time. First, no problem had strongly captivated me; second, any problem worth investigating would need my full-time attention; third, I was in no particular hurry to produce another paper, since I had already written more than two dozen and published most of them in reputable journals. Some of my UCSF colleagues thought my focus on scrapie was a passing fancy and that a more promising career could be found in studies of glutamate metabolism of the nervous system. Glutamate is the most abundant amino acid in the brain, I was a neurologist, and I had had considerable training in the study of enzymes that convert glutamate into glutamine and vice versa. But I could not be dissuaded.

Evidently I believed that the scrapie problem was relatively easy before I started my investigations. Neurology chairman Bob Fishman's recollections provide an interesting assessment:

He was in the residency finishing up, and I said, "Stan, what do you want to do?" He said, "Well, you know, I spent a lot of time doing protein chemistry with Stadtman." To paraphrase what he said, in so many words, he wanted to work on the scrapie agent, because, number one, he thought it created these interesting degenerative diseases and, number two—and I've tweaked him about this—he said, "It ought to be an easy problem to solve. It's just a problem in protein chemistry."[5]

Looking back at this turning point in my career, I'm baffled by my determination on the one hand and rather impressed on the other. I was sufficiently insecure that I hungered for success; I was motivated, ambitious, driven. I had grown up in a family with very little financial means or social status, so whatever I wanted from life I had to make happen. My parents' abundant unconditional love was surely instrumental in creating a personality that did not buckle under criticism—which, as it turned out, would come from many corners as the astonishing properties of the prion were discovered.

Interlude: Viruses, Genes, and Proteins

Viruses are tiny infectious agents that can be seen only in an electron microscope. Viruses are one-tenth to one-one-hundredth the size of bacterial cells such as *Staph* and *Strep,* both of which can be seen in the light microscope. Every virus contains an inner core of genes. A gene is the smallest unit of genetic material—most genes encode proteins, which are the action molecules of life. Genes are composed of the nucleic acid DNA, except for the genes of some viruses. Such viruses possess RNA genes instead of DNA; both polio and the AIDS virus HIV are RNA viruses, while herpes is a DNA virus. The RNA or DNA genomes of viruses reside within the tightly packed core of the virus, which also contains proteins encoded by the viral genome. These viral proteins participate in the production of new copies of the virus. Viral nucleic acid contains all the instructions necessary for the production of nascent viruses. The exterior surface of a virus is composed of viral proteins and, sometimes, of proteins and lipids co-opted from the surface of the host cells.

DNA is often referred to as the "genetic material of life." Most genomes consist of double helices of DNA, forming twisting ladders whose rungs are composed of four different bases—adenine (A), thymine (T), guanine (G), and cytosine (C)—arranged in varying sequence to form genes. In going from the DNA gene to protein, there is an intermediate step called messenger RNA. After a messenger ribonucleic acid (or mRNA, as it's usually called) is assembled, it is used to instruct the protein synthesis machinery precisely how protein should be created.

The order of the bases, which is specified by the mRNA, creates a code that directs the assembly of twenty different amino acids into chains called proteins. Different combinations of three bases specify which amino acid should be added next in a growing chain. For example, the sequence "CCA" encodes the amino acid proline. Deciphering the genetic code as well as the colinearity between nucleic acids and proteins was a major step in unraveling biological information that is passed from parent to offspring.

The chains of amino acids that form proteins are assembled inside cells beginning from one end called the N-terminus. When the last amino acid is added, that end is designated the C-terminus. By convention, when scientists describe the beginning of a protein, they place the N-terminus on the left and the C-terminus on the right. Proteins are the essence of life; they perform many different functions and are the major components of all cells. Some proteins called enzymes speed up chemical reactions that are required for life, while others act as receptors and still others function as scaffolds. The biological activity of a protein requires that the amino acid chain fold into a compact shape called a conformation. Shape is everything to a protein—denaturing a protein stimulates its unfolding into an inactive chain of amino acids. As the shape of a protein disappears, so does its biological activity. Sometimes, a protein can be refolded to regain its biological activity.

3

A Plethora of Theories

By the time I had begun contemplating a study of the scrapie agent at UCSF in the early 1970s, laboratory investigations into its nature had been greatly enabled by transmission of the disease from sheep or goats into laboratory mice. This watershed moment was first reported in 1961, by Richard Chandler of the British Agricultural Research Council's facility at Compton, in Berkshire, England.[1] Subsequent transmission of the scrapie agent from mouse to mouse provided the best available experimental system.

The challenge was to apply this system to develop a scheme that would separate the scrapie agent from other substances in the brain, such as nucleic acids, proteins, and fat. Brain homogenates from infected mice had to be manipulated so as to enrich particular fractions for these substances. But so little was known about the physical characteristics of the mysterious scrapie agent that hundreds of fractions would have to undergo titration measurements.

Since I was not going to see any firm results for a year after starting the experiments, I needed to set up at least a dozen or so possible purification schemes. If each protocol took two or three days to implement, then I could have all the samples ready for titration measurements in about a month. Assuming that each of the dozen schemes generated twenty-five samples, I would have three hundred samples to measure. So far, so good, but now came the hard part—I needed sixty mice to determine the number of scrapie agents in each of the three hundred samples. I got a sick feeling in my stomach when I realized that three hundred multiplied by sixty equals eighteen thousand mice! I would have to find enough money to buy eighteen thousand mice and, worse, pay to house them for a year. I would also need to hire a couple of highly skilled technical people who could examine the mice daily for neurological signs of scrapie. In those days, a mouse cost a couple of dollars, plus a nickel a day to house, so the bill for a year would be about $450,000, including the salaries for the two technicians.

And there was one more problem: Preparing the samples and injecting them into healthy mice would take at least three months, and nine more months had to pass before the results from the first titration measurements would begin to appear. To avoid being idle during those nine months, I needed another $1.5 million, in order to have four staggered sets of eighteen thousand mice running all the time. Needless to say, I was never going to be able to raise enough money to maintain a daily census of seventy-two thousand mice! Such a war chest was pie in the sky. Where was I going to find $2 million for each of the next ten years?—I thought that at least ten years would be required to prepare highly purified samples of the scrapie agent. At most, I might be able to get $50,000 annually from an NIH research grant for animal care. Whenever I discussed my plans with my colleagues, they told me I had to develop a faster, cheaper, easier procedure for quantifying the scrapie agent.

Interlude: My Initial Protocol for Measuring the Scrapie Agent

Brains of sick mice were homogenized, diluted, and injected into weanlings, which developed signs of neurological dysfunction about 150 days later. To ascertain the number of scrapie agents in a mouse brain, a series of measurements called a titration was performed. After grinding up a mouse brain in a salt solution to create a 10 percent homogenate, we created a 1 percent homogenate by adding nine parts of saline. By repeating this dilution step, we made a 0.1 percent homogenate. When such dilutions were performed serially, repeated six more times, the scrapie agent was diluted a billionfold. Typically, six mice would be injected with each dilution and examined twice a week for neurological dysfunction. Those mice receiving several hundred thousand scrapie agents developed disease in about 150 days, while those inoculated with one scrapie agent took more than 250 days to manifest illness. A final tally was taken after a year, to allow all the mice replicating the scrapie agent sufficient time to become ill.

Such measurements required fifty to sixty mice (six mice for each of eight to ten dilutions) to determine the number of scrapie agents in a single sample. In the study of viruses and bacteria, such dilution assays have been used for many years and are often referred to as "endpoint titrations." In my own mouse studies, the endpoint was the last dilution able to produce scrapie infection.

I was determined to find a way through this jungle of problems. Conceptually, I was unintimidated, even though the logistics were daunting. To make matters worse, the mouse titration method was imprecise and unable to measure small increases in the purity of the scrapie agent. To show reproducible results using this system, the enrichment for the scrapie agent had to be at least a hundredfold. Building a protocol to isolate the scrapie agent from a multitude of mouse-brain contaminants was to prove far more difficult than I had envisioned. If my fascination with the scrapie problem had not blinded me, I might never have started the work. A careful assessment would have convinced me that obtaining adequate funding for such a risky enterprise was far too difficult. Moreover, I would have realized that receiving tenure was most unlikely if my efforts did not result in publishable papers after a year or so. The hubris of youth was all that propelled me forward.

Despite the tedious, slow, and imprecise mouse dilution assays, Tikvah Alper had managed to discover the unprecedented resistance of scrapie infectivity to inactivation by X-rays and ultraviolet light, so why shouldn't I achieve a modicum of success? Since a single X-ray photon should be sufficient to kill a single scrapie agent, Alper was able to calculate the minimal size of the agent; she estimated that it was less than one-one-hundredth the size of a typical virus.[2] To extend her ultraviolet light irradiation studies, Alper used ultraviolet light of different wavelengths, in collaboration with Raymond Latarjet in Paris. In 1970, they reported that at 237 nanometers the scrapie agent was six times more susceptible to inactivation than at either 260 or 280 nm. Alper and Latarjet thought their results argued against both a nucleic acid and a protein, since nucleic acids absorb light maximally at 260 nanometers—one nm is equal to one-billionth of a meter—and proteins at 280 nm.[3]

While the results of Alper's ultraviolet inactivation studies were provocative, there were many possible interpretations—caveats abounded.

For example, a thick protein-lipid-sugar coat might shield a viral genome from much of the damaging ultraviolet light. Or the nucleic acid genome of the putative "scrapie virus" might be composed of modified bases (the building blocks of nucleic acid) that were resistant to inactivation by ultraviolet light at the standard germicidal wavelength of 260 nanometers. Still another possibility was that the damaged nucleic acid core of the scrapie virus was easily repaired and therefore looked much more resistant to ultraviolet light than it was.

My view of Alper's X-ray studies was similar, in that there were multiple explanations for the small target size she had calculated. Perhaps a protein coat laced with metals deflected the X-ray photons, thus shielding a viral genome. As with ultraviolet damage, special bases might make the nucleic acid core resistant to inactivation by X-rays—or the X-ray-injured nucleic acid core might be easily repaired and therefore look much more resistant than it was. With my limited knowledge of radiation biology, I thought all these explanations were worth considering. I also decided not to dismiss the most startling interpretation: *All the data might be pointing to an infectious particle devoid of nucleic acid and thus with no apparent way to replicate.* If this were true, then it would be worth an enormous effort to decipher the structure. An important discovery might emerge from this work.

Quite naturally, Alper's findings prompted a wealth of speculation. By the early 1970s, hypotheses about the scrapie agent outnumbered its investigators. A fascinating array of structural hypotheses was offered to explain the unusual features of scrapie and its infectious agent. Among the earliest hypotheses was the notion that scrapie was a muscle disease caused by the parasite *Sarcosporidia.* After the successful experimental transmission of scrapie from sheep to sheep in the late 1930s, however, the virus hypothesis became popular. Some investigators proposed that the scrapie agent might be an RNA provirus copied into a DNA genome. (A provirus is a virus that inserts itself into the genome of a host cell and is thus passively replicated.) The "scrapie provirus" might be exceedingly stable, which would explain the unusual properties found by Alper and her colleagues. Others proposed replicating sugar chains—that is, a nucleic acid surrounded by a protective coat of sugar polymers—as likely structures. Yet others proposed a cell membrane fragment. RNA tumor viruses, for example, bud from the surface of cells and take on part of the cell membrane as a protective coat; might the scrapie agent be an amorphous fragment of cell membrane that somehow acquired the properties of an infectious pathogen? Some investigators even proposed a protein, and one prescient mathematician suggested several mechanisms by which a protein might replicate itself.[4]

Since the most stable of all known viruses are those with relatively small DNA cores, such a virus was proposed as a likely structure for the scrapie agent. (Of the small DNA viruses, the parvovirus is the best known; in the United States, it is the most common cause of fatal infection in young dogs.) Another proposal was a DNA "subvirus," a substance smaller than a virus but with viral properties, controlled by a "transmissible linkage substance"; such a proposition made no sense to me, so I dismissed it at the outset. It seemed irrational and full of gobbledygook.

Unable to gain any insight into the nature of the scrapie, kuru, and CJD agents, kuru czar Carleton Gajdusek proposed the term "unconventional virus," but he provided no structural details about how these unconventional "virions" differed from conventional viral particles.[5] Iain Pattison, of the British Agricultural Research Council's Compton research institute, grasped the inanity of Gajdusek's label in a memoir: "The fourth decade of my association with scrapie ended in 1978, with the causal agent still obscure, and virologists as adamant as ever that theirs was the only worthwhile point of view. To explain findings that did not fit with a virus hypothesis, they had rechristened the causal agent an 'unconventional virus.' Use of this ingenious cover-up made 'virus' meaningless—for is not a cottage an unconventional castle?"[6]

In 1973 the Swiss-American plant pathologist Theodor O. Diener suggested another possible identity for the scrapie agent. A half-dozen years earlier, he had discovered small, naked RNA molecules that infect plants and mimic viruses, and he had named them viroids. Virologists had shown that the nucleic acid core of some viruses is, by itself, infectious under artificial conditions in the laboratory, but these large viral genomes cannot survive in the natural world without a protective protein coat. Unlike viruses, viroids have no protein coat. Their small size was at least consistent with Alper's X-ray experiments that showed the scrapie agent to be quite small.

What Alper and her colleagues had not done was to define the molecular components of the scrapie agent. Only purification of the agent—that is, separation from cellular debris and other substances in mouse-brain homogenates—would lead to an unambiguous explanation of their results. Evidently, they did not know how to perform the necessary studies. Their lapse saved the problem for me: Purification of the scrapie agent would be my singular focus.

I viewed the problem from the vantage point of a chemist. To identify the agent's composition, I needed to begin by removing impurities that might be distorting the interpretation of Alper's results. Certain impurities might render the scrapie agent resistant to inactivation by ultraviolet and X-ray radiation, or make it appear smaller, or camouflage a buried nucleic acid. Once pure preparations were available, I could determine the agent's molecular components.

There had already been many attempts to purify the scrapie agent, each designed to test one or another of the hypotheses put forth about its structure. Each was predicated on the assumption that achieving purity would be relatively easy, like the purification of many viruses. I felt that some of these biochemical studies were conceptually flawed and their execution poor. In my opinion, many critical aspects of the work had not been done systematically and lacked the discipline characteristic of studies performed by well-trained chemists. I was prepared to work on the problem for a long time. Only by progressively enriching a homogenate from some mouse organ for infectivity could I expect to see the scrapie agent ultimately emerge.

Where to begin? I had to formulate a plan that would yield publishable data, but the logistics and cost were daunting. No reasonable person wanted to perform assays that required sixty mice and one year to obtain a single reliable data point. Equally imperative, the critical experiments needed to be repeated not once but twice before publication. I had to be sure that others were going to be able to confirm my work. The scientific process depends on others being able to reproduce independently those experiments that herald meaningful advances. And what if you set up a laboratory to do such work and the answers you obtained were trivial? After all, the vast majority of scientific investigations yield very little new information, and even then most of the new information is incremental. Big leaps forward are rare.

My First Job

By the summer of 1974 I had been working in other people's laboratories off and on for eleven years, and I yearned to be in charge of my own creative endeavors. I began to look for a position where I could pursue my newly acquired passion. After considering a move to Duke University, I decided to remain at UCSF. Bob Fishman's invitation to join the Neurology Department as an assistant professor was generous: He assigned the second-largest laboratory in the department to me, consisting of two rooms on the seventh floor of the Health Sciences East Building. Their previous occupant was the neurochemist Elizabeth Roboz-Einstein, who had recently retired. She was married to Hans Einstein, a professor of engineering at University of California, Berkeley, and the son of Albert Einstein. I could thus revel in an exceptional pedigree for my lab space, but I preferred to ignore the relationship, fearing that it might raise unobtainable expectations. Along with my salary and the salary for a technician, Fishman gave me a modest supply budget and funds for a few small pieces of equipment. I also had a small animal room near my laboratory that could hold three racks of mouse cages, but because of the high cost of animal care, I could afford to use only one.

Now all I needed to do was persuade the NIH to send lots of funding to UCSF for scrapie studies. Experimental science requires plenty of money. Without costly laboratory supplies and equipment and attractive salaries for smart assistants, modern science cannot be done. Having decided on a problem, I needed funds to support the work. It all seemed reasonable to me: Write a proposal for the NIH that describes earlier failures, then lay out a systematic plan for success. But I was not prepared for the lack of sympathetic understanding among some of the reviewers of my grant application.

Although my applications for an NIH research career-development award and an NIH research grant on glutamate metabolism had been funded, my application for support to study the scrapie agent was poorly received. It had been assigned to the NIH's Virology Study Section, whose members thought scrapie was caused by a slow virus and were skeptical of the chemical approach I had outlined. They considered the study of scrapie to be a virological problem, which should be approached quite differently from the research plan I proposed. Studies of human and animal viruses in mice and rats had progressed slowly for many years, until a way had been found to grow viruses in cultured cells. In such cells, viruses multiplied to high levels and proved to be an ideal source for vaccines, such as the one Jonas Salk and coworkers developed for polio. Lawns of cultured cells in petri dishes could also be used to count viruses by determining the number of foci of either dead or rapidly growing cells. Many virologists thought that such approaches ought to be used in trying to determine the molecular structure of the scrapie agent.

To my disappointment, the application not only was "not funded," it was "disapproved," the reviewers arguing that I had no training in virology and was thus unqualified to undertake studies on the "slow virus" causing scrapie. They couldn't see why I might be successful where others with proper training in virology had failed. The reviewers also noted that I had no expertise in scrapie research. Though these were reasonable criticisms, I was angry. I could not understand why these virologists did not share my vision and could not see the logic of my approach. I considered my lack of experience in virology an advantage—bringing a fresh perspective to a so-far-intractable problem—but to respond to their criticisms I spent three weeks in the summer of 1975 at the Cold Spring Harbor Laboratory on Long Island studying animal viruses.

Collaboration with William Hadlow

I also initiated a collaboration with William Hadlow, at the NIH's Rocky Mountain Laboratories in Hamilton, Montana. Hadlow was probably the first veterinarian to undertake a residency in human neuropathology; he had trained at the University of Minnesota under James Dawson, who was later famous for his description of subacute sclerosing panencephalitis (SSPE), a slowly progressive brain disease also known as Dawson's encephalitis and caused by the measles virus. More to the point, Hadlow had once worked on scrapie at the United Kingdom's Compton Institute. On July 3, 1959, he had taken the train to London and gone to an exhibit on the pathology of kuru sponsored by the Burroughs-Wellcome Company. When he looked at the numerous small round empty spaces or vacuoles in brain sections from kuru patients, he found the vacuolation pattern indistinguishable from that caused by scrapie. Struck by this similarity, he published a short paper in the British journal *Lancet* entitled "Scrapie and Kuru," in which he emphasized how closely the

Figure 7   William Hadlow made the leap from scrapie to kuru in 1959.

neuropathologic changes in the brains of patients with kuru resembled those in sheep and goats with scrapie.[7] Hadlow argued, based on scrapie studies, that the best way to pin down the disease's cause was to inoculate chimpanzees with brain extracts from dead kuru victims and hold the animals for many years to see whether they developed the disease. By the time of our collaboration, he had returned to the Rocky Mountain Laboratories and set up a program to study the pathogenesis of scrapie in sheep (figure 7).[8]

Having bolstered my credentials, I resubmitted my grant application to the NIH, accompanied by a rebuttal. To my delight, the grant was funded. With my first NIH grant on scrapie, my work achieved a modest degree of legitimacy among my peers. But while the grant gave me funds to pay a technician and buy supplies, the virologists in the NIH study section cut my animal-care budget severely, deleting the funds I needed to purchase thousands of mice and pay for their daily care. How they thought I was going to do the work I had outlined in the proposal was unclear. But the grant at least gave me the necessary footing to raise money from private donors and foundations. My initial foundation support came from the Howard Hughes Medical Institute, and my first private support came from a kind and gentle businessman, Clarence Heller, one of the potential donors whom Julius Krevans, the dean of the UCSF Medical School, had approached. Krevans usually prefaced such requests by saying, "Stan's research is risky, but the payoff could be big and the implications of his work are enormous." He would emphasize the uncertain outcome of my efforts, while noting that UCSF had built such a successful biomedical research enterprise that "we can afford to support a few high-risk projects." I would cringe inwardly while I listened to these presentations, though I knew he was trying to be fair with prospective donors. But I did not consider the work all that "high-risk," and I was beginning to believe that the scrapie agent might prove to be different, in kind and not in degree, from all known infectious agents. Moreover, I was sure that whatever we found for scrapie would apply to the CJD agent; there was no reason to think otherwise.

The severe budget cuts in my NIH grant were to remain a major problem. Over the next two years, I wrote two supplemental applications to the NIH requesting additional funds for animal care, neither of which was approved.

4

The Scrapie Race

ow I found myself in a scientific race, albeit one that began at an exceedingly slow pace. Unfortunately, the lack of systematic investigations meant that I needed to start from scratch. The race would soon all but take over my life.

During the previous decade, Bill Hadlow and his Montana colleague Carl Eklund had performed a large number of experiments on the pathogenesis of scrapie in sheep and goats, as well as some studies on experimental scrapie in mice. While neither of them had tried to define the molecular structure of the scrapie agent, Eklund, a virologist who thought like a chemist, was enthusiastic about my proposed systematic purification; he had witnessed numerous failed experiments designed to unravel the nature of the scrapie agent using a hit-or-miss approach. Many of these experimental results were published, but the findings rarely provided insight into what the next step might be. Instead of an incremental step forward in characterizing the nature of the scrapie agent, the experimental design was generally geared toward solving the mystery in one fell swoop. Such experiments had little chance of success, since the number of variables was so large.

Once I had devised a starting point, I would break the research into fundable stages for which support could be obtained—aware that if I made a mistake at the outset, there might not be enough time to find my way out of the cul-de-sac. To begin, I had to choose a source of the scrapie agent—a tissue with which to start the purification process. Eklund and Hadlow, as well as Richard Kimberlin at the Compton Institute, had measured the replication of the scrapie agent in spleens and brains of mice. Both groups found that the levels rose rapidly in spleen tissue immediately after inoculation but plateaued after about a month. In contrast, the levels in brain tissue reached a maximum after about three months and were several hundred times greater than those in spleen. At first glance, the brain seemed like a superior source, but its high fat content made this organ more difficult to process than the spleen. The adherent fat caused the brain's proteins and nucleic acids to smear across many fractions.

Hadlow and Eklund agreed to inoculate fifty mice and harvest their spleens forty days later. Since there were no published data on the behavior of the scrapie agent in brain or spleen homogenates in the centrifuge, and neither Hadlow nor Eklund had any unpublished studies, I designed a set of centrifugation experiments. A centrifuge is a precision instrument in which samples in special tubes can rotate at speeds up to 100,000 revolutions per minute. Conditions can be defined under which the particles of interest either remain suspended in the relatively clear liquid known as the supernatant or collect in the sediment at the test tube's bottom, known as the pellet. I wanted to learn the conditions under which the scrapie agent remained in the supernatant fraction and those when it pelleted. Such an inglorious study might be a prudent starting point. Since each assay required sixty mice, we sought every economy possible, and after much discussion we decided to measure only the scrapie agent in the supernatant fraction, ignoring both the pellet and the interface, thus reducing the number of assays by two-thirds.

Before embarking on the experiment, I bought some normal mice and collected their spleens and brains. With these uninfected organs, I produced homogenates, fractionated them using the centrifuge, and measured the levels of several enzymes and of RNA in the supernatant fractions. After several modifications of the procedure, I was able to obtain smooth curves showing that the enzymes and RNA in the supernatant decreased as the centrifugation speed and time increased.[1] For the real experiment, I ground up some of Hadlow's infected mouse spleens in saline to create homogenates. Portions were centrifuged at carefully chosen speeds for precise periods, and each supernatant was frozen and sent to Hadlow in Montana for serial dilution; each of ten dilutions was then inoculated into groups of six mice.

•

Now the clock began to tick: We expected supernatant samples containing high levels of the scrapie agent to produce disease in four to five months. I guessed that such samples were likely to come from supernatant fractions obtained when the centrifugation speed was low and the time of centrifugation was relatively brief. Would we get meaningful data? Would there be a clear relationship between the level of the scrapie agent and the conditions of centrifugation? We had to wait a year for the complete story, but after four or five months we might get a glimpse. Still, the study was a gamble. It would probably not yield any breakthroughs, but it might lay a solid foundation for constructing effective purification protocols. Could such conservative experiments—confined solely to the scrapie agent's sedimentation properties—yield data that were publishable? If not, I was in trouble, but without such information I didn't know how to begin purifying the agent. It was not obvious that I would be able to package the results from such studies into interesting papers, and I was nervous about the outcome.

After six months, the supernatant samples that had been centrifuged briefly and at low speeds began to cause neurological disease in mice. After another six months, the news was good: We had smooth curves and significant changes. The smooth curves indicated that increasing the centrifugation speed and time caused progressively more of the scrapie agent to be pelleted. The ten thousand–fold reduction in the concentration of the scrapie agent from the supernatant was sufficiently large that it was likely to be reproducible and possibly useful in devising a purification protocol.

But there was bad news, too: The gentle slope of the curve indicated that the scrapie agent existed in many different sizes; this meant that centrifugation could not, by itself, easily yield a substantial enrichment in scrapie infectivity. Developing an effective purification scheme was not going to be easy.

After two years of NIH support, I had to write a renewal application. In it, I described the progress that Hadlow, Eklund, and I had made—I was pleased with it, although there had been no breakthroughs—and outlined the work that needed to be done next. The NIH responded by stating that four members of the Virology Study Section would come to UCSF for a site visit to review the work. Since the Study Section questioned neither Hadlow's nor Eklund's expertise or their part in our collaboration, they were spared this inquiry.

As the day approached, I grew increasingly nervous, and I prepared diligently for their questioning. Typically, site visitors meet the night before, to review their assessments and decide among themselves on critical questions. Since nearly nine months had elapsed from the time I began to write my renewal application, I thought it would be prudent to give each of them a notebook with my latest data. I was ready for an hour or two of questions; to my surprise, they spent nearly six hours that day cross-examining me. Six months passed before I learned the outcome: My renewal application had been disapproved. I was distraught—and unnerved by the virologists' critique of the recent data I had so thoughtfully provided, in which they accused me of misleading them. They had asked why the control traces from a brain-fractionation study were the same for two experimental studies conducted nine months apart. I had explained that since numerous controls had always given similar profiles, I did not repeat them each time I worked with scrapie samples. My budget was tight, and repeating the controls before I had a meaningful result from the scrapie samples seemed unnecessary. Once we developed a useful protocol for enriching the scrapie agent, I told them, we planned to perform the proper control studies prior to publication.

However, the virologists were dissatisfied; in fact, they came close to accusing me of scientific misconduct. The allegation was so serious that I was uncomfortable asking the Virology Study Section to judge my science again. Indeed, one person from the study section told me that some of the members were so unsympathetic to my work that they had quashed my earlier supplemental applications for animal-care support. Their minds were closed, he said, and he recommended that I find another source of funds. This experience taught me a valuable lesson: Preliminary data have no place in grant applications or site visits; you need to be sure the data are reproducible, the controls properly done, and the interpretation well developed. On the rare occasion when I've deviated from this approach, I've been sorry.

Fortunately, I had support from a large NIH Program Project grant that Dick Baringer had managed to secure a year earlier, so I could afford to let the matter drop. Baringer and some other senior neurovirologists at UCSF had assembled an application to the NIH on multiple sclerosis and virus infections of the nervous system. Since they believed scrapie was caused by a slow virus, they invited me and Hadlow to join the effort (a fortuitous circumstance not without its irony, as things turned out). This time the review committee, composed of virologists, neurologists, and biochemists, evinced considerable enthusiasm for my work, and I got all the funding I had requested. Much later, I learned that a neurologist on the site-visit team had stayed up late into the night trying to persuade one of the virologists to be supportive.

By 1977, almost three years after starting our studies, we finally had enough data for a small paper on the scrapie agent, with smooth curves showing changes in infectivity over a ten thousand–fold range as a function of a physical parameter. It was an authoritative answer to the question I had posed to numerous scrapie researchers: Under what centrifugation conditions will the scrapie agent sediment?

I asked Robley Williams at UC Berkeley, an expert on the physical characteristics of viruses, to consider communicating our paper to the *Proceedings of the National Academy of Sciences*. After reading it, he called me to say that there were problems and asked if I would come to his office to discuss them. When I arrived, his manner was forbidding. "I'm not going to send this paper to the *Proceedings*," he said, "but you may be able to find someone else to do it." My brain sent a telegram to my gut telling it that all was not well. Could I save the situation or was it a bust? Hoping for the best, I asked him if I could explain to him what we had been doing. Williams had recently reached the mandatory retirement age and his emeritus status gave him some extra time, so he was willing to sit and listen to me. When I finished, he said, "I like it. That's pretty good work. I'll find a couple of referees and get it reviewed." Apparently the paper had not conveyed the message I was able to present verbally. There was no guarantee that it would be accepted, he warned, and I would have to wait for the verdicts of the referees. If they were positive, he would communicate it. A few weeks later, he forwarded their comments and some additional suggestions for improving the paper, and when I sent him the rewrite he communicated it. I was relieved to have a publication on the scrapie agent in such a highly respected, widely read journal.[2] It was crucial to sustaining my grant support.

Based on the centrifugation results, I developed a scheme for purifying the scrapie agent from both spleen and brain: it began with homogenization of the tissue followed by a brief low-speed centrifugation to pellet large debris and leave the scrapie agent in the supernatant. This step was followed by a high-speed centrifugation that sedimented more than 99 percent of the agent. After resuspending the pellet in saline buffer, we added a detergent to extract unwanted fat and repeated the centrifugation sequence. After removing the supernatant, with its numerous detergent-fat complexes, we used high-speed centrifugation to collect the pellet, which we dispersed in a saline-detergent solution.

Once I was convinced that spleen tissue offered no advantage over brain in terms of ease of purification, I changed my approach, since the scrapie agent was several hundred times more abundant in brain. Next, I began to think about ways to extend our purification of the scrapie agent. I realized that scaling up our preparations was probably critical, and I guessed that we might need to increase the number of mouse brains in a single preparation from fifty to five hundred or even twenty-five hundred. At the same time, I began to contemplate the biosafety problems that such large preparations of an infectious agent could present. I was concerned that once we started to scale up our preparations, one of my laboratory staff might become infected. Might the scrapie agent cause disease in people, even though no one had connected scrapie and CJD in several epidemiological studies? After all, the animal and human spongiform encephalopathies shared many features, as Hadlow had pointed out in his paper on scrapie and kuru. Because we needed to centrifuge large quantities of the scrapie agent, I wanted to enclose the centrifuge in a biosafety cabinet that could withstand a rotor failure at high speed.

Safety Cabinets and Biocontainment

Fortunately I found an expert on biocontainment just across the Bay, at the Oakland Naval Bioscience Laboratory: Mark Chatigny, an engineer with expertise in the design of high-containment facilities for studying deadly pathogens. Laminar airflow hoods were beginning to be widely used in biomedical research. The Baker Company of Kennebunk, Maine, built the best hoods, so I called the company president and discussed our need for centrifuge enclosures. Chatigny and I wanted to explode an ultra-high-speed rotor inside the enclosure to test its integrity. With a "yes" from Baker, we approached the Beckman Instrument Company to explode a rotor, explaining that we needed to find out whether the biosafety cabinet containing the centrifuge was strong enough to contain infectious agents during a rotor failure.

The cabinet was shipped from Kennebunk to a Beckman plant in Palo Alto, where an ultracentrifuge was placed inside it. Chatigny filled tubes with bacteria, inserted them into the rotor, and placed agar plates throughout the cabinet and the surrounding room to capture any bacteria that might escape when the rotor failed. A spectacular series of tests ensued. The fifth run proved to be a winner. Just as the rotor reached forty thousand rpm, it exploded with a deafening roar. The biosafety cabinet shook but stayed intact, and we all cheered. Mark collected the agar plates and took them to incubators, where any bacteria that had escaped could grow. After a few days, he scored the plates and found that whereas some of the bacteria had made their way out of the centrifuge into the biosafety cabinet, none had escaped into the room.[3] I was delighted by the results, and we immediately ordered two Beckman centrifuges enclosed within Baker biosafety cabinets for our newly built biocontainment facility.

Though our centrifugation curves allowed us to identify the optimal conditions in which most of the debris in the homogenates was sedimented and the majority of the scrapie agent remained in the supernatant on a small scale, we were now ready to try it on much larger preparations. Using the information from our initial study, I added detergent to bind as much of the remaining fat as possible in the supernatant just as we had done before, and centrifuged the sample again, but this time at a high speed, so as to collect the scrapie agent in the pellet.

So far, so good, but now I wanted to refine our somewhat crude approach even further. The next step would be to increase the centrifuge's resolution by using a sugar gradient, atop which would be the dispersed pellet containing the scrapie agent in the saline-detergent mixture. The most syrupy layer would be at the bottom of the tube, and on it would be stacked layers of progressively less density. This gradient would cause small particles to collect at the top and large particles at the bottom; most important, medium-size particles would move to the middle. By modifying the concentrations of sucrose and the speed and time of centrifugation, we could calculate the size of the particles collecting in the middle. As it happened—and much to my horror—the scrapie agent did not segregate into fractions of differing infectivity at top, bottom, and middle. Instead, infectivity was spread everywhere, with the small particles proving as infective as the medium-size and large particles! This was a nightmare. Nature was not about to give up her scrapie secrets easily.

There were two interpretations: One was that the small particles were an artifact of the centrifuging process and the big ones represented the real scrapie agent; the other was that the small particles were the agent and the bigger ones were aggregates of it. Most scrapie researchers, and virtually all virologists, favored the first interpretation; after all, viruses are relatively large particles, and the scrapie agent was thought to be caused by a slow virus that had yet to be isolated. I thought the second interpretation was more reasonable, because it was consistent with Tikvah Alper's radiation studies, which had demonstrated that the scrapie agent was roughly one-one–hundredth the size of a typical virus.

Membrane Fragments and Viruslike Particles

Back when Alper was performing her provocative radiation studies, Richard Kimberlin and his supervisor, Gordon Hunter, were carrying out a series of biochemical studies at the Compton Research Institute (figure 8). Kimberlin was a tall young man with a shock of thick red hair and a personality cheerful, enthusiastic, and self-assured. His most distinctive facial features were his large nose and warm eyes. His soft-spoken demeanor was at times disarming, as was his engaging conversation. Hunter was very different—a somewhat apathetic but kind and gentle, middle-aged man with a full head of gray hair. His lack of success in deciphering the molecular structure of the scrapie agent weighed heavily on him. By the early 1970s, seemingly exhausted and frustrated, Hunter concluded that the scrapie agent was a membrane fragment, with no regular, recognizable structure—an amorphous complex of lipids and other undefined stuff.[4]

Hunter and Kimberlin had no idea how such a membrane fragment might replicate except by recruiting other membranes. Most important, they argued that the scrapie agent could not be purified, because disruption of the lipid-rich membranes by detergents would destroy its infectivity.[5] Had I believed that the scrapie agent was a membrane complex and as such could not be purified, I would have abandoned my purification efforts. But I did not believe it—I thought their studies were not well formulated and their approach haphazard. Or maybe I just wanted this idea to be wrong so I could pursue the research project to which I had become so committed. I was grateful that their lack of success left the door ajar for me. Most scrapie researchers found the membrane hypothesis as unappealing

Figure 8   Alan Dickinson (top left), Richard Kimberlin (top right), and Richard Marsh, along with other scientists not shown, investigated the molecular structure of the scrapie agent. Marsh, a veterinarian, is shown examining a patient.

as I did. Without a viable alternative, many of them, especially those with virological training, were eager to return to the idea that a virus causes scrapie. They needed only to explain why Tikvah Alper had misinterpreted her data. For instance, Carleton Gajdusek and his postdoc Robert Rohwer at the NIH argued vehemently that Alper's findings were easily compatible with a virus, noting that the extreme resistance of the scrapie agent to radiation damage was explained by an efficient repair mechanism.[6]

By 1978 my idealism was beginning to tarnish. My career as a scientist seemed unlikely to flourish if I continued using mouse titrations to measure the scrapie agent. No matter how efficiently I designed those experiments, I had to wait a year for the results, and I began to despair of producing real advances in the field. After all, only thirty sets of sequential experiments could be completed in a three-decade-long career, and I had learned firsthand that purification of the scrapie agent was a biochemical nightmare. The agent did not behave in a way that was going to make its purification straightforward. Clever tricks were needed; developing a purification scheme was going to require many, many titration assays in order to negotiate a path through the fog.

Moreover, my life had become increasingly stressful as others claimed to have successfully isolated the scrapie agent. Shortly after I had decided to pursue its purification, a report appeared describing a small transmissible agent that caused scrapie and multiple sclerosis. The data generated quite a stir, but no one could reproduce the findings. A year later, another report appeared, claiming that the scrapie agent was a tiny virus that had a spherical shape when viewed in the electron microscope. These spheres were about twelve nanometers in diameter, making them about half the size of the smallest known virus, and their size seemed consistent with Tikvah Alper's findings. We tried to repeat the work, but I concluded that the studies were so poorly executed that the chances of their being correct were slim. Richard Kimberlin put much more effort into repeating these studies and eventually showed that the spherical particles were ferritin, a complex of proteins that carries iron and is thus readily visualized in the electron microscope.[7]

Using the electron microscope, investigators had reported numerous sightings of the putative scrapie virus in the brains of scrapie-infected animals before the ferritin mishap. Among the earliest reports was one by Joe Gibbs in 1968, in which he identified collections of spherical particles about 23 nm in diameter in mouse brain.[8] These particles were of a respectable size for a virus—they were similar in appearance to polioviruses—but their relationship to scrapie was never established. When Dick Baringer also found these particles in the brains of scrapie-infected mice ten years later, he became excited, as did many virologists with whom he shared the photographs.[9] I could not understand their enthusiasm. If these relatively large spherical particles caused scrapie, then how could anyone possibly reconcile them with Alper's findings? Deep inside, I wanted the scrapie agent to differ from a virus. By that time, I had put so much effort into my studies that I did not want to find some trivial answer. This was an emotional response.

The longer I worked in the field, the more nerve-wracking these claims from others became. While my dedication to my research program steadily expanded and the allure of the problem escalated, my anxiety rose with each report. I found myself in the middle of a competition I had never intended to enter. But now that I was there, I did not want to finish second. Over time, I began to appreciate how dangerous it was to participate in such a contest. Being in a scientific race brings a pervasive pressure to interpret the data in support of your hypothesis. I constantly had to fight against this temptation.

The idealism that pervades scientific investigation fades when you enter a full-fledged competition. Often I found myself feeling possessive of the problem and angry with others who did not see me as its rightful owner. It was as though I believed that my efforts had created a homestead— that I had acquired squatter's rights. That someone else would determine the molecular structure of the scrapie agent first was the grimmest of all prospects. A few pages in a reputable journal can render another scientist's years of toil virtually worthless.

Virinos and Viroids

The competition produced some strange behavior. The idea of a small nucleic acid appealed to the veterinary geneticist Alan Dickinson of the University of Edinburgh, whose studies had demonstrated the existence of strains of the scrapie agent that seemed to breed true, thus suggesting the presence of nucleic acid (see figure 8). Dickinson introduced the term "virino" in 1978:

If the recent experimental results . . . are correct implicating DNA as a necessary component of the infective unit of scrapie, then an appropriate name for this class of agent would be "virino," which (by analogy with neutrinos) are small, immunologically neutral particles with high penetration properties but needing special criteria to detect their presence.[10]

Later he and others modified the definition to "viroid-like DNA complexed with host proteins."[11]

Dickinson was passionate about his work, and some of his studies were clever and well executed. He was a thin man of average height and was forthright, a bit arrogant, and somewhat humorless. His rigidity and deeply held conviction that he was right and others were wrong did not endear him to his colleagues. The extremely long time frames of his experiments and his lack of interest in shortening the time were his downfall. Moreover, his severe views of other people's limited intelligence were never masked, and he was so opinionated that it was difficult to have an earnest conversation with him. My first encounter with him was in the summer of 1976, at the Third International Congress of Virology in Madrid. Many of the people working on scrapie, kuru, and CJD were gathered for lunch on the patio of a restaurant near the congress hall. During lunch, I remarked, "What the scrapie field really needs is a rapid assay." Upon hearing me, Dickinson roared, "That would be the worst possible situation! Then we would have two hundred people in the field. This field is the right size now." His campaign to keep scrapie mysterious continued for another decade, until he clashed with the British funding authorities and lost his job.

But Dickinson was not alone in what I considered to be undisciplined approaches to uncovering the nature of the scrapie agent. The scrapie race was threatening to cut corners that should not have been cut. In 1978 Richard Marsh, a professor of veterinary science at the University of Wisconsin, using a procedure similar to that employed in the isolation of viroids, reported the preparation of samples from the brains of hamsters with scrapie which could be inactivated by digestion with the enzyme DNase—which specifically destroys DNA (see figure 8).[12] Neither RNase nor protease inactivated the scrapie agent in this purified fraction. For more than two years, most investigators working on scrapie thought that a small DNA molecule similar to the infectious RNA comprising viroids was the cause of scrapie. Shortly after Marsh's work appeared in *Nature,* I learned that a graduate student working in Kimberlin's laboratory had identified a small DNA molecule in fractions prepared from brains of mice with scrapie, but not from controls. Presumably this was the viroid-like DNA discovered by Marsh and his coworkers.

To check whether Marsh's report was correct, I set out once again to reproduce the work of others. The procedure described to us by a technician at the University of California, Riverside, who had worked for Joe Semancik, one of Marsh's collaborators, seemed straightforward. There was a genuine elegance to the simplicity of the isolation procedure, markedly different from the convoluted protocol that had produced the little spheres that were eventually shown to be ferritin. I remember thinking, "If the isolation of the scrapie agent is that easy, then I don't belong in science." Was my systematic approach too tedious? Were these more imaginative ways of unlocking the secrets of nature the hallmarks of good science? But after more than a year of work and many tens of thousands of dollars expended, we could not repeat Marsh's results. We had no evidence for a viroidlike DNA that caused scrapie. I felt conflicted: On the one hand, I had spent a lot of time and effort showing nothing more than that someone else was wrong; on the other, I was happy that the answer to "What causes scrapie?" had not yet been found. I still had a chance to cross the finish line first.

Needing to resolve our negative findings, I called Marsh and told him I had been unable to repeat his studies. I wanted his help. "Call Joe Semancik in Riverside," he said. "He'll help you."

I was unprepared for what happened next. I phoned Riverside and spoke to Semancik. "I would be most appreciative if you'd come to San Francisco and examine our data," I said. "You could tell us where we've gone wrong."

Semancik seemed uninterested in helping me. He told me that he was far too busy to leave Riverside and that I should keep repeating the experiments. He urged me to persist and eventually I would get them to work.

"We've performed these studies three times," I protested, "and we always get the same answer. We have no evidence for a DNA viroid causing scrapie."

In response to my complaint, he pressed me to continue repeating my studies, and they would eventually work.

Trying to keep my cool, I suggested he send his technician to San Francisco and let her examine our data. After he rejected this request, I took a deep breath and said, "I'd like to bring our data to Riverside and let you or your assistant examine them."

Again, he told me that he and his technician were far too busy to spend time with me trying to resolve this discrepancy. He just couldn't find time to see me. I was disturbed and frustrated. Why was Semancik unwilling to examine our findings? Why wasn't he upset that I couldn't reproduce his data after spending the better part of a year trying to repeat their work? I phoned Marsh again and asked him if I could bring my data to him.

"Stan, I don't know how to do these studies," Marsh said. "The purification was done in Riverside. The purified samples were then sent to me so I could determine the scrapie agent titers in hamsters."

I was taken aback to learn that Marsh hadn't tried to transfer some of the purification work from Riverside to Madison. After all, if the work was correct, they had made an important discovery. Marsh needed to know whether their results were reproducible; otherwise, he was heading down a blind alley.

At this point, I decided we should publish a paper describing our findings. After we did so,[13] neither Marsh nor Semancik challenged our conclusions. To my annoyance, neither of them wrote a retraction of their own data. Shortly thereafter, I learned that the DNA that Kimberlin's graduate student claimed to have identified was a fraud. The student could reproduce his work only when everyone else in the laboratory was having a few beers at the local pub after working hours. This debacle served as an excuse to close down scrapie research at Compton and to fire Gordon Hunter. Curiously, Kimberlin emerged from all of this unscathed and set up a new research program under Alan Dickinson, in the Neuropathogenesis Unit at the University of Edinburgh.

I was still in the race. As far as I could tell, no one had figured out the composition of the scrapie agent. Hypotheses remained abundant. The field was wide open. All that was needed were great ideas and superbly designed experiments. But I had no great ideas. I still didn't know how to design the clever experiment that would give the "Aha!" answer. Where was the eureka? Why couldn't I see it?

5

Dr. America and the Trembling Cannibals

Player number one in the competition I had joined was undoubtedly Carleton Gajdusek of the NIH, guru of kuru and proponent of the "unconventional virus." His and Joe Gibbs's successful transmission of CJD to a chimpanzee in 1968, demonstrating that it was an infectious disease, had its origins in the Eastern Highlands of New Guinea, where he eventually set up camp in longer and longer hegiras from the NIH, and where I was to have an enlightening visit with him. Gajdusek, who died in 2008, was an engaging, forceful, insecure individual whose rambunctious personality remained unaltered throughout the thirty-four years of our acquaintance. His severe Slovakian features were distinctive, and his dark brown straight hair parted on the left partially covered his wrinkled forehead, which was punctuated with deep grooves. He was addicted to hyperbole and he loved hearing his own baritone voice; he was driven to dominate every conversation in any room he found himself. Once before a lecture, I asked him, "Carleton, are you getting nervous in anticipation of your talk?" He answered swiftly, "No, I am only nervous when I am not talking." A masterful speaker, he was famous for his three- or four-hour-long lectures; he could keep an audience spellbound. He also polarized people: Some of his colleagues were devoted and remained close to him for decades; others, annoyed by his wild exaggerations and his rambling monologues, found him difficult and judged his flight of ideas and occasional inaccuracies inconsistent with rigorous scientific investigation.

New Guinea is the second-largest island in the world. Its eastern half, which also encompasses several smaller Melanesian islands, is Papua New Guinea. There, small tribes of natives live in a tropical rain forest ten thousand feet above sea level. *Homo sapiens* has been living on the island of New Guinea for more than fifty thousand years, and during that time more than seven hundred distinct languages have evolved, reflecting the extraordinary isolation of the tribes. The western half is called Western New Guinea and is part of Indonesia; Papua New Guinea was a German colony until it was assigned to Australia by the League of Nations after World War I, and after the Second World War, Australians began to explore the mountainous interior. In the early 1950s patrols in the Eastern Highlands reported on the practice of sorcery among the Fore (pronounced "Foray") people, and a patrol officer named John McArthur identified a trembling state known as kuru, which was ascribed to sorcery by the natives but which McArthur thought to be of psychological origin.

In 1956 Vincent Zigas was assigned as an Australian medical officer to the Fore region as part of his mandatory national service after medical school. He was impressed by the trembling and the difficulty walking exhibited by many young Fore women. As a physician, he wondered whether the tremors and unsteady gait suggested some form of Parkinson's disease. Anthropological studies were no help, since the Fore ascribed the trembling to sorcery. In the 1950s kuru was the most common cause of death in young adult women in the Fore region. Puzzled by the affliction, Zigas was delighted when Gajdusek, a pediatrician as well as a virologist, arrived in March 1957 in search of interesting research opportunities (figure 9). Gajdusek's medical training and his connection with NIH associate director Joseph Smadel, giving him access to many resources and laboratory studies, made him an ideal colleague. Zigas and Gajdusek decided to carry out an epidemiological study. They defined the geographical boundaries of the illness and its gender and age distribution. Kuru was confined to the Fore and bordering tribal areas. The largest afflicted group consisted of young women, followed by children of both sexes, but with girls outnumbering boys two to one. The few adult male victims were generally older than the women. These epidemiological findings resembled no known disease in the annals of Western medicine; when Zigas and Gajdusek published the first medical reports of the

Figure 9   The New Guinea–NIH connection. Carleton Gajdusek (left) and Vin Zigas, shown attending to a boy with kuru, established the experimental transmission of kuru and Creutzfeldt-Jakob disease to apes and monkeys.

illness in 1957, they introduced the Fore word "kuru" into the scientific literature.[1]

To characterize kuru further, Gajdusek and Zigas performed autopsies and harvested pieces of virtually every organ. But their attempts to identify a toxin or infectious agent causing kuru were unsuccessful; equally unproductive were attempts to identify a metabolic or genetic cause. Their studies revealed no abnormalities in any organs except the brain. On visual inspection, the brain tissue looked normal, neither swollen nor shrunken, but under the microscope numerous holes, or vacuoles, were seen—particularly in the cerebellum, which controls coordinated movements.

Hadlow and Transmission of Kuru to Chimps

This brain tissue was to provide the first important clue to kuru's exact nature. Gajdusek sent some of it to the neuropathologist Elisabeth

Beck at the Maudsley Hospital in London. Her ability to produce informative sections of brain tissue for microscopic examination was outstanding. But the scientist who made the connection between kuru and scrapie was an unlikely observer from afar. Bill Hadlow, on sabbatical at the Compton Institute, spotted the resemblance when he saw Beck's brain sections at the 1959 London exhibit on the pathology of kuru.

Almost seven years elapsed between Hadlow's 1959 proposal in the *Lancet* for experimental studies on the transmission of kuru using non-human primates and Gibbs and Gajdusek's pioneering report of such a transmission. How could such an important experiment have been so long delayed? Hadlow's argument seems at first to have occupied a rather low priority in Carleton Gajdusek's mind. Perhaps he had grown possessive of the field—or perhaps he was uninterested in carrying out Hadlow's experiment because he was not a coauthor of Hadlow's *Lancet* paper. In 1961, Joe Smadel, to whom Gajdusek reported at the NIH, tried to recruit both Hadlow and Richard Johnson, a Johns Hopkins neurologist and expert in viruses that attack the nervous system, to work on a kuru transmission project, but neither was interested in becoming a "chimpanzee doctor." In the spring of 1962, Smadel persuaded Joe Gibbs to take the job. Gibbs was a modest, clean-shaven, devout Catholic bachelor. His gentle blue eyes peered through rimless glasses that were framed by a bald head adorned along the sides with short white hair. Joe was a thin man of average height who let his professional life be dominated by Gajdusek's overwhelming personality.

While Gibbs would care for the apes and monkeys, someone else had to obtain the brain samples from dead kuru patients and ship them from the jungle to civilization, where they could be injected into chimpanzees. That task fell to Michael Alpers, a young Australian physician who had become excited about Hadlow's proposal while he was a medical student in Perth. Alpers moved to New Guinea in October 1961 and began developing the resources needed for transmission studies. By March 1962 he had decided to build a house in Waisa, a small Fore village. That month, he met Gajdusek in Okapa, an Australian patrol outpost about a one-day walk north of Waisa. A couple of days later, Gajdusek visited Alpers in Waisa, and they decided to collaborate on the transmission studies.

In September, Alpers performed his first autopsy—on a thirteen-year-old boy named Eiro. In March 1963 he performed a second, on an eleven-year-old girl named Kigea, who had lain almost motionless for six weeks before dying (neurologists describe such people as "locked in"). The brain samples were frozen and shipped to the NIH in Bethesda. In August 1963 a small piece of Kigea's brain was homogenized and a 10 percent suspension was injected into the brain of a chimpanzee called Daisey. The next month, another chimpanzee, Georgette, was inoculated with a 10 percent suspension of homogenate prepared from Eiro's brain, and in February 1964 a third chimp, Joanne, was inoculated with a suspension from a twenty-three-year-old woman named Kabuinampa, who had died the preceding December.

Part of the delay had been caused by the difficulty in obtaining chimpanzees for the experiments, Alpers recalled when we spoke over lunch in London in September 2007. "Lots of people wanted us to use monkeys, but I wanted to use chimpanzees initially," he told me. "What if our experiments with monkeys had failed? I wanted to follow Hadlow's suggestion and use the animal most closely related to humans."

The chimps were still healthy in the spring of 1965, but by midsummer Georgette had started to show difficulties with coordination. ("I told Joe Gibbs, 'She looks like she has kuru,'" Alpers remembers.) Soon afterward, Daisey became ill, and Gajdusek returned to the NIH from his field studies in Papua New Guinea to monitor the illness. In October, Georgette was sacrificed and an autopsy performed. Alpers and Gibbs removed the brain for postmortem examination and gave it to Elisabeth Beck, who took specimens to London for neuropathological examination. By early February 1966, with Daisey dead and Joanne ill but holding her own, the pathology studies were completed, and Beck sent a telegram stating, "Neuropathology in chimp indistinguishable from kuru." The next day, Alpers, Gajdusek, and Gibbs wrote their paper on the transmission of kuru and submitted it to *Nature* for review. It had all happened just as Hadlow had envisioned seven years earlier.

Cannibalism

While writing their landmark paper, the authors began to speculate about how kuru was transmitted among the Fore and their neighbors. Since cannibalism was common among these people, the possibility that it featured in the transmission of kuru became a subject of considerable discussion once the first transmission studies were published. The ritualistic aspects of cannibalism among the Fore have been debated over many years. Some Fore thought that the soul resides in the brain, so eating the brain of a dead relative was the way to give the deceased eternal life. In the early 1950s anthropologists had described the Fore's complex death rituals, including cannibalism, but by the end of that decade cannibalism had begun to disappear as the Fore came under pressure from the Australian government and local missionaries. By the time the first transmissions of kuru to chimpanzees were completed, the number of patients with kuru had sharply decreased.[2]

In recounting his studies on cannibalism in the early 1960s, the anthropologist Robert Glasse wrote:

> The strongest evidence suggesting an association between kuru and cannibalism can be found in the consumption patterns of kuru victims. If it is true that kuru was initially a disease of women only, as the inhabitants claim, then the absence of the disease among adult men today or at least its low incidence, might have a cultural explanation. Throughout the kuru region, when Europeans first entered the area, the great majority of women were cannibals. . . . Children consumed any food given to them by their mothers, including human flesh. The behavior of male cannibals varied by region and from individual to individual. In the South Fore, cannibalism among adult men was extremely rare, if it existed at all, partly because of the belief that cannibalism robbed a man of his vitality, thus making him vulnerable to enemy arrows. . . .
>
> After the funeral, the victim's maternal kin dismembered the corpse. They took the body to a sugar cane garden, where they butchered it with a bamboo knife and stone ax. They first removed the hands and feet then cut open the arms and legs, stripping out the muscles. They then opened the chest and belly to remove the viscera, taking special care to excise the gall bladder, which, if ruptured, would ruin the taste of the meat. They next severed the head and fractured the skull to remove the brain. Little was wasted: marrow, viscera, and brain were all cooked and eaten. Sometimes even the bones were pulverized and eaten with green vegetables.
>
> Two methods of cooking were employed. The body could be cooked whole in an earth oven, in much the same way that pig is

cooked. This method was employed in the North Fore. In the South, many women preferred to steam small pieces of flesh in a bamboo tube with salt, ginger, and leafy vegetables. The North Fore, in addition, sometimes buried a body for several days to improve the flavor, considering maggots an extra delicacy.[3]

While cannibalism presented a plausible explanation for the transmission of kuru from one person to another, initial attempts to transmit the disease to chimps orally were negative.[4] To counter this negative finding, Gajdusek argued that when the women and children opened the skulls of kuru victims, they were exposed to high levels of the infectious agent. Since they rarely washed, because there was no running water in their villages, the kuru agent remained on their hands for prolonged periods of time; so it may well have entered the body through the oral mucosa when they put their hands in their mouths, or through conjunctival membranes whenever they rubbed their eyes. Another route was thought to be sores on the hands or forearms. In 1980, however, Gibbs and Gajdusek reported that one of two monkeys fed brain and other tissues from chimps with kuru over a five-day period developed central nervous system dysfunction after three years.[5]

The similarities between the microscopic neuropathology of kuru and Jakob-Creutzfeldt disease (as it was then called) had been noted in the same year, 1959, that Hadlow published his *Lancet* paper. That observation lay dormant until the experimental transmission of kuru to chimpanzees. Soon afterward, Gajdusek and Gibbs injected a brain extract from a victim of CJD into a chimp, and two years later they reported on the transmission of the disease. Gajdusek, Gibbs, and their coworkers were on a roll. Not only had they transmitted an exotic neurological disease found in the highland jungles of New Guinea, but they had also transmitted a brain disease described almost fifty years earlier by Alfons Jakob and Hans Creutzfeldt. The transmission of two degenerative brain diseases of unknown etiology was an important accomplishment by any measure.

In their report of the transmission of Jakob-Creutzfeldt disease to a chimpanzee in 1968, there was a transposition of the names Jakob and Creutzfeldt. I was puzzled, because numerous neurologists and pathologists had argued that the disease should be called either Jakob's disease or Jakob-Creutzfeldt disease, but hardly anyone called it Creutzfeldt-Jakob disease. Alfons Jakob wrote a paper in 1921 describing several cases of progressive dementia with widespread neuronal loss in the brain.[6] A year earlier, Hans Creutzfeldt had described the brain of a woman who died after a prolonged series of seizures.[7] He found widespread vacuolation, but some medical scientists believe that his patient died of a seizure disorder complicated by hypoxic brain damage and doubt that she had what became known as CJD. So why did the names Jakob and Creutzfeldt become flipped in an important report on this disease, regarding the transmission to a chimpanzee? Twenty years passed before I found the answer, in conversation with Gibbs. "When I was writing my first paper on the transmission of Jakob-Creutzfeldt disease to an ape," he told me, "I wanted to rename the disease 'Gibbs disease.' I didn't think this would be acceptable to the scientific and medical communities, so I decided to reverse the names, because my first name is Clarence and my middle name is Joseph and my initials are C. J." Thus did the disease become known as Creutzfeldt-Jakob disease, or CJD for short—a case of mind-boggling scientific mischief.[8]

Kuru Patients Up Close

From the beginning of my fascination with the scrapie agent and its relation to human disease, I knew I would have to examine people dying of kuru. To learn about a disease, physicians must see the patients. You cannot otherwise appreciate its details and nuances, and you need to see many patients over time and watch the disease progress or remit. So in the spring of 1975, I asked Carleton Gajdusek whether he would be willing to arrange for me to examine the Fore. Although his reply was affirmative, no definite invitation to join him in New Guinea was forthcoming, despite my repeated requests.

I had first met Gajdusek the preceding summer, while visiting Richard Kimberlin and Gordon Hunter in Compton. Gajdusek's visit prompted Hunter to organize a dinner at his home, and he kindly included me. Gajdusek, it appeared, loved to monopolize conversation. He talked nonstop after we sat down to eat, and when everyone else had finished eating, he was just starting. The peas presented a real problem; they kept falling off his fork. Each time someone tried to say something, Gajdusek would lower his fork, his peas would plummet onto the plate, and he would raise his voice so that everyone else stopped speaking. Tiring of this mono-logue, eventually we all fell quiet so Gajdusek could eat his peas.

After dinner, I talked to him about the possibility of organizing a section in his NIH laboratory on the biochemistry of the scrapie agent. Gajdusek had the resources to mount a serious effort on defining the molecular composition and architecture of the scrapie, kuru, and CJD agents. But he explained to me that he did not want to enlarge his labora-tory, because that would increase his responsibilities at the NIH and thus necessitate his spending more time there and less in the field. He was under constant pressure, he said, to diminish his time in New Guinea and other exotic places with primitive cultures. I have reflected several times on what would have happened if Gajdusek had embraced my idea of moving back to the NIH. Though I was disappointed at the time, his reluctance was a stroke of luck.

In early October 1976 I knew my proposed trip to New Guinea under his auspices would be further delayed. I was driving to the San Francisco airport to catch an airplane to Missoula, Montana, for a meeting with Bill Hadlow and Carl Eklund when over the radio came the announce-ment of the award of the 1976 Nobel Prize in physiology or medicine to Baruch Bloomberg, a specialist in hepatitis B virus, and Carleton Gaj-dusek "for their discoveries concerning new mechanisms for the origin and dissemination of infectious diseases." I was surprised that Hadlow, Joe Gibbs, and/or Michael Alpers had not shared the prize with Gajdusek. When I arrived in Missoula, Hadlow was there to meet the airplane and drive me to the Rocky Mountain Laboratories. He seemed to accept the award of the prize to Gajdusek with no ill feeling. Even if it had bothered him, I imagined, Hadlow's stoic dignity would not have allowed him to admit it.

My eagerness to go to New Guinea remained steadfast, but I received no help from Gajdusek in arranging a trip. Not surprisingly, he was being pulled in a hundred different directions: People all over the world were clamoring to hear him speak about kuru and CJD. Receiving the Nobel Prize had made him an even more popular lecturer.

Dick Johnson, now a professor of neurology at Johns Hopkins, had visited the kuru region in the mid-1960s, since he had a long-standing interest in slow viruses, particularly those that attacked the brain. John-son was one of those magical figures in neurology—charismatic, self-confident, happy, jocular, and engaging. A shock of thick prematurely gray hair parted on the left crowning an infectious smile punctuated Dick's delightful personality. His encyclopedic knowledge of neurovirology provided a wonderful platform from which to launch an endless array of entertaining stories about many of the temperamental scientists who figured prominently in the field. Since Gajdusek was so preoccupied, Dick suggested that I contact Alpers, who was now the director of the Medical Institute in Goroka. Soon thereafter, Alpers arranged my visit. In the fall of 1978 I flew from San Francisco to Auckland and on to Port Moresby, where I changed planes for Goroka. Alpers was in China when I arrived, but his wife met me, and I stayed with her and their four children for a few days before traveling by truck to Okapa and on to Purosa, in the heart of the Fore region, accompanied by several Fore workers from the Medical Institute. From Purosa, we hiked for a couple of hours to the village of Agakamatasa, which sits at the southern edge of the kuru region, high above the Lamari River (figure 10). Much to my surprise, Gajdusek was there to greet me—he was to be my host after all. Only later did I learn that the Fore workers from the institute worked for him, since NIH funds supported many of the institute's activities, including continuing surveillance studies of kuru.

The villagers referred to Gajdusek as Doctor America. He was kind to them, and they liked him immensely. Each morning, people from the surrounding area would come to the village to be treated by him for a variety of ailments. Much of Gajdusek's magic emanated from the antibiotics he injected them with to treat their respiratory infections, cellulitis, or abscesses, all of which are common among primitive peoples with poor hygiene and frequent exposure to smoke-filled huts. While I was not surprised by the admiration this engendered, I was unprepared to listen to Gajdusek speak to the natives in Melanesian pidgin. He had told me how fluent he was in many languages, including Fore, but to the Fore he spoke pidgin.

Gajdusek's quarters consisted of a large circular hut twenty-five feet in diameter and nearly twenty feet high at the apex. The frame was constructed largely of bamboo stems lashed together with reeds; the walls and roof were thatched grass. There was a small kitchen and a dining area with a table and four generous wooden chairs. A ladder led to a sleeping loft.

Figure 10 High above the Lamari river (top) sits the village of Agakamatasa (middle) at the southern edge of the kuru region. Patient with advanced kuru exhibited frontal lobe release signs: snout and sucking reflexes (bottom) and hand grasps (not shown).

Each evening, the natives served us dinner, consisting mainly of taro, sweet potatoes, corn, and other local vegetables. One evening we had wild boar for dinner; it had been roasted for hours in a deep pit and though quite well done, it had a succulent flavor. This was clearly a festive occasion, and I felt properly honored. Had I visited the Fore twenty-five years earlier, no doubt similar cooking pits would have contained the brains of dead relatives wrapped in large fern leaves or stuffed into bamboo logs.

At these dinners in Doctor America's hut, I was allowed one sentence, and from that point on all the talking was done by Gajdusek, who would launch into a lecture on the seven hundred languages of New Guinea, or the early exploration of the Fore region, or the value of fiberglass tape. I found his sweeping statements and exaggerations maddening. Exhausted from listening, I would eventually excuse myself, climb up into the loft, and stretch out on the grass floor with a couple of blankets. As I began to doze off, I could see Gajdusek still at table, holding forth as though I were still sitting across from him. Eventually, he would realize that I had decamped and would terminate his monologue, climb up the ladder, and go to sleep across the room.

In the morning, I usually set out on a long hike, in pursuit of peace and quiet, in preparation for another evening of bombast. As the week wore on, my tolerance for Gajdusek's hyperbole and my respect for his scientific skills lessened—though frequently I found myself thinking that something was backward here. I had it all wrong. I should be thinking the opposite way around. I should feel privileged to be in the kuru region with the man who opened up this fascinating area of medical investigation. I should be grateful that Gajdusek was giving me so much of his time. I was a nobody. I had yet to accomplish anything substantial in this field.

After a week in Agakamatasa, I set out with two guides, Anua and Auyano, to find Fore people who were dying of kuru. I came to enjoy my guides' company. They were warm, kind, and gentle men, and, like many of the Fore, highly intelligent, though they had had little schooling. Anua was just under five feet tall with delicate facial features and narrow shoulders. Auyano was slightly larger with coarser features and large brown eyes set deep in their orbits; most prominent was a large hole in his nasal septum, where he had worn a pig bone in his youth. Both of them liked to laugh, and we had no trouble communicating in pidgin.

The beauty of my surroundings overwhelmed me. At nearly ten thousand feet above sea level, there were no mosquitoes—what a blessing! The mountainous jungle was different from any landscape I had ever encountered, its steep hillsides covered with tall, slender beech trees and many smaller trees, including oak, ficus, bamboo, and pandanas. More than fifty feet above us was a thin tropical canopy soaking up the sunlight. Everywhere were ancient ferns, small shrubs, and occasional orchids and rhododendrons, covering the mountainsides in a soft, thick carpet. Women and children tended terraced vegetable gardens in the numerous hamlets. Some of the paths were treacherous, especially after a heavy rain; thick, slippery mud was everywhere. My footwear was not ideal for slogging through the mud, which did not stick to the heavily calloused bare feet of my Fore guides but clung maddeningly to the Vibram soles of my hiking boots. At times, I felt I was skiing on small toboggans. This sort of ambulation was manageable when I was descending a gentle slope, but walking uphill was hopeless. Over the next two weeks, we visited many villages. I recall fondly the sugar cane and the short, squat bananas, which I never managed to find in the United States.

On this expedition, I was able to examine eight patients afflicted with kuru. Some were in the early stages of the disease and managed to walk with the aid of a stick. Their most profound neurological deficits involved coordination. Like a drunk, a typical patient could not walk a narrow line by placing the toes of one foot against the heel of the other. Her balance was so compromised that she was unable to stand on one foot. When sitting, she was unable to place one heel on the opposite knee and smoothly move it down the shin. When I asked her to move her index finger from her nose to touch my index finger, she showed a pronounced intention tremor—that is, a tremor produced during voluntary movement. The patients also had difficulty with rapid alternating movements of the hands. Such neurological difficulties with coordination, called limb ataxia, occur in malfunction of the lateral lobes of the cerebellum.

In more advanced cases, patients were unable to walk even with the assistance of a stick. Some had so much trouble with their balance that they could stand only when grasping a heavy pole firmly planted in the ground. The swaying from side to side was so pronounced in a few cases that they could not even sit up without the aid of such a pole. This swaying occurs when the medial lobe of the cerebellum can no longer receive and send signals that keep the trunk erect. As the trunk starts to lean in one direction or another, the cerebellum is signaled that a correction is needed. In a normal person, the correction signals are so fine and incremental that you see no change in the position of the trunk. But in a person with midline cerebellar dysfunction, when the body drifts off center, the correction is coarse and this malfunctioning process is repeated over and over again; the alternating sequence is called truncal ataxia, also referred to as titubation.

Besides damage to the cerebellum, the patients with late-stage kuru showed signs of widespread damage to their frontal lobes. These patients exhibited a grasping reflex when I gently touched their palms, and snout and rooting reflexes when I touched the skin around their lips (see figure 10). Such primitive reflexes are seen in babies but disappear after the age of two; they reappear when there is damage to both frontal lobes. Snout and rooting reflexes are commonly seen in patients with Alzheimer's disease and other dementias, including CJD.

The initial medical descriptions of kuru by Vincent Zigas and Gajdusek had focused on the tremors. This led to analogies with Parkinson's disease, in which tremor is a major feature. Although both Zigas and Gajdusek were physicians, neither was a neurologist, so they must be forgiven for their inaccurate clinical assessments. Later, when Richard Hornabrook, a neurologist from New Zealand, visited the Fore region, he readily differentiated the intention tremor of kuru from the resting tremor of Parkinson's disease. He described kuru as a cerebellar disorder and documented the progressive limb and truncal ataxia.[9] Zigas and Gajdusek, as well as Hornabrook and many anthropologists, noted episodes of uncontrollable laughter displayed by some kuru patients, but I did not witness such behavior. I did notice that some of the patients with early kuru seemed appropriately depressed, since they knew their condition would worsen over the next six to nine months and end in death.

In 1980, I made a second visit to the kuru region, by which time the epidemic had all but disappeared because of the Australian government's efforts to stamp out cannibalism. However, I was able to examine seven more patients with kuru, expanding my clinical experience to a total of fifteen. When I returned from New Guinea, I brought one of Anua's sons, named Sena, with me. He spent the night at my home, and the next day I sent him on to Bethesda, where he was to be adopted by Gajdusek and educated in Maryland public schools. By that time, Gajdusek had adopted more than forty Melanesian children. When I listened to him describe these children, I wondered whether this was an experiment in the adaptation of children raised in primitive cultures to Western society. My feelings vacillated between admiration at the opportunity he was giving them and dismay at the cruelty of separating them from their parents. As it happened, some of Gajdusek's children were to have more to contend with than homesickness.[10]

6

The Battle for Tenure

ack home at UCSF, I had become frustrated. My collaborative work with Bill Hadlow was proceeding much too slowly. We had managed to publish only three papers on the behavior of the scrapie agent during centrifugation. I was beginning to realize that my approach to understanding its composition was not realistic. Something better had to happen or my grant support would disappear. My fantasies of success were fading, my productivity was unimpressive, and my chances of tenure were declining.

I was tired of waiting for answers while the biological sciences were exploding with new technologies, including gene cloning. I wanted to apply these extraordinary new techniques to the study of the scrapie agent—to look for a gene that might somehow be involved in its replication. How could we ever understand what causes scrapie by performing mouse-titration assays, given the time this approach required?

In 1975 Richard Marsh and Richard Kimberlin had reported an isolate of the scrapie agent that in a large dose caused disease in Syrian hamsters in about seventy days[1]—this was about half the time necessary for mice to manifest clinical signs of scrapie. When I asked Hadlow about switching from mice to hamsters, he was unenthusiastic; hamsters required larger cages and needed more bedding and food. Another problem to solve! Then came a call in January 1978 from Kenneth Sell, the scientific director of the NIH's Institute of Allergy and Infectious Diseases, who said that the Rocky Mountain Laboratories' Board of Scientific Counselors had evaluated Hadlow's program and that I could no longer collaborate with him. Sell told me that when the counselors had asked Hadlow about his work with me, he replied that he didn't really understand the experiments and they should get the details from me. From this statement, they concluded, not unreasonably, that we did not have a real collaboration and that I was using Hadlow's resources to do my own work—thus violating, in their opinion, the border between the intramural NIH program of the Rocky Mountain Laboratories and its extramural, competitive research grants program. They believed, in other words, that I was doing an end run, using "intramural" resources to do "extramural" research at UCSF. Clearly I was, because I couldn't raise enough money to pay for all my animal expenses, which were beyond the bounds of normal biology research budgets. In terms of research costs, scrapie was the high-energy physics of biology.

Needless to say, I was upset, particularly because I had not been at the NIH review to answer their questions. Sell remarked to me that Hadlow had known about the site visit months in advance. Why hadn't he asked me to be present, or at least to review our experiments with him in advance? When I asked Hadlow to fight this decision, he was unwilling to do so. He had a fine scientific reputation but no stomach for protesting my case to the NIH authorities. I found his passivity maddening. Reflecting back on the situation, I wonder whether Carl Eklund's death a few months earlier might have played a role in Hadlow's apparent lack of interest in pursuing our collaboration. Eklund's training in virology may have provided a better framework for understanding what I was trying to accomplish than Hadlow's morphological background. Whereas Eklund was retired and thus had had no responsibility for the money involved in measuring the scrapie agent in my samples, Hadlow was accountable and had to justify how his research funds were spent.

I remain grateful to Bill Hadlow for having helped to launch the project. Moreover, terminating the studies at the Rocky Mountain Laboratories was a key turning point—it proved to be unanticipated good luck. In an effort to be helpful, Ken Sell suggested that I move to the Montana labs to continue my work. Sell was a tall man with broad shoulders and a loud, self-assured voice. He was seriously overweight, which may have accounted for his death at age sixty-five due to complications of his diabetes. His enthusiasm for scientific investigation was refreshing, and he was particularly kind and helpful to me. When I told Ken that I appreciated the offer but didn't want to move, he proposed a short-term contract to ease the transition and gave me funds to set up an animal facility in the Bay Area.

Earlier attempts to convince the veterinarian at UCSF's animal-care facility that I needed my own animal colony had failed. At one point, I had asked Hadlow to come to San Francisco to talk to him. Since they were both veterinarians, I thought Hadlow's intervention would be persuasive. It wasn't. I wanted to operate my own facility to ensure that the caretakers did what was needed, but I could not afford the high per diem rates the university charged. Besides caretakers, UCSF wanted me to pay for a hierarchy of administrators. I couldn't possibly raise the money the university demanded, and unless I managed to solve this problem, the research would grind to a halt, so I was forced to look elsewhere. After an agonizing search, I was able to establish a small animal colony at the Naval Bioscience Laboratory in the Oakland Navy Yard with the NIH funds from Ken Sell.

Now I could begin my studies of scrapie in hamsters. My technical assistants, Darlene Groth (figure 11) and Patricia Cochran, inoculated them with an extract of a hamster brain sent to us by Dick Marsh. Sure enough, to our delight, they developed signs of scrapie about seventy days later. When we used progressively diluted brain extracts, we noticed that the higher the dilution, the longer the time from inoculation to the onset of neurological dysfunction.

The cost of caring for a hamster was five times that of a mouse, and I decided we might save money by turning the situation upside down. Why not use the length of the incubation time to measure the number of scrapie agents in the inoculum? After collecting data for half a dozen samples, in which sixty hamsters were used to measure the number of scrapie agents in each of the samples by titrations, we constructed graphs showing that the number of scrapie agents correlated with the length of the incubation time. There was an inverse relationship between the number of days and the number of infectious agents. Next, we tested this relationship by measuring the behavior of the hamster scrapie agent in the centrifuge, as we had done previously with the mouse scrapie agent. The hamster scrapie agent data mimicked that of the mice. The new numbers were so good that I became convinced we could measure the amount of scrapie agent in a sample by inoculating only four hamsters. Reducing the number of animals from sixty to four and the time to score the test from a year to seventy

Figure 11 Developing a new assay for the scrapie agent. Darlene Groth (above), Patricia Cochran, and I showed that the number of scrapie agents correlated well with the length of the incubation time.

days accelerated our work by a factor of almost eighty. We could do in one year what previously would have required nearly a century! Our new measurement procedure changed the landscape of scrapie research. Over the next three years, we performed more assays on the scrapie agent than had been done in the entire previous history of studies on scrapie.

With a daily census of sixteen hundred hamsters, we were able to assay the scrapie agent in four hundred samples simultaneously. Since the hamster scrapie agent assays required less than 120 days to complete, three rounds of assays allowed us to determine the scrapie agent in twelve hundred samples annually. As described in chapter 2, using mouse titration assays to establish the number of scrapie agents in twelve hundred samples would have required a census of seventy-two thousand mice. While we were never able to raise the $2 million needed to care for all those mice, we raised enough to maintain a sizable hamster colony. When the hamsters showed signs of neurological dysfunction, they were sacrificed and their brains were rapidly frozen in liquid nitrogen and stored until they were needed in some of our protocols.

Unfortunately, in 1979 we were forced to leave the Naval Bioscience Laboratory, because the navy veterinarian became frightened of scrapie. Mark Chatigny, who had arranged for us to house our hamsters there, was furious at the veterinarian, who was irrational about the risk of scrapie being transmitted from animals to people. Now I was in another quandary. Chatigny suggested that I explore the Letterman Army Research Institute in the Presidio. This seemed like a good idea, but I didn't know where to begin. Perhaps I ought to approach a member of Congress? On my next trip to Washington, I visited Representative Phillip Burton, San Francisco's congressman, and described my plight to one of his staff members. Somehow I struck a sympathetic chord: The staff person saw my work as a cause that might enhance Burton's image. I followed up with a letter to Representative Burton asking for his help. I told him how the army could facilitate this important work, which was still being funded largely by NIH research grants. He replied that he would do what he could. For several months nothing happened. While waiting for Burton to act, I was fortunate enough to find space for my animals at the vivarium of UC Berkeley. Then one day I got a call from an army colonel:

"Is this Dr. Prusiner?"

"Yes."

"I have been told by Harold Brown, the secretary of defense, to make you happy."

"That's great!" Instantaneously, I began dreaming that the U.S. Army was going to pay for all my animal expenses—manna from heaven. Somehow Congressman Burton had convinced the Defense Department, so recently beleaguered by the Vietnam War protests, that helping me would be great public relations.

"Sir, I understand that you want to use the animal facility of the Letterman Army Research Institute for your work."

"That's correct. How do we make that happen?"

"Sir, I'll tell the commanding officer, and you should visit him. I'm sure you'll want to see the facilities."

"Thank you. And please thank Secretary Brown." I replied.

Well, how about them apples! After negotiating a high-security entrance to the animal facility at Letterman, I was escorted on a mammoth freight elevator to the fourth floor by a sergeant and introduced to a major, who gave me a tour. He showed me the biggest autoclave for sterilization I had ever seen; it was big enough to engulf a medium-size truck. The cage-washing facilities were similarly impressive. But then the disappointment began. There were many rooms for rabbits, dogs, and nonhuman primates but relatively few rooms for housing rodents, so where were all my hamsters going to live?

After the tour, I discussed with the major the cost of using the army's animal Taj Mahal. The per diem rates were so high I nearly choked. And it was not clear that he understood that he was supposed to make me happy; rather, his intent appeared to be to discourage me so that I would leave him alone. A few days later, I wrote to Congressman Burton and Secretary Brown, thanking them for their efforts and letting them know that while I thought Letterman was a wonderful place, I had been able to arrange animal space at the University of California, Berkeley, facility that was sufficient for my current needs.

With my animals safe in the comparatively modest confines of the Berkeley vivarium, I thought I was doing reasonably well. My work was moving forward, and I was regularly publishing papers in refereed journals. Still, some of the neurology faculty at UCSF thought I was wasting my time. With so many neurological diseases untreatable, surely I could spend my efforts devising therapies instead of studying a little-known disease of sheep. Many neurologists viewed my project as a "crazy idea," and most basic science faculty discouraged graduate students from working with me—my project was simply too fanciful. They thought it was never going to produce meaningful findings. As a consequence, research assistants and a few brave postdoctoral fellows staffed my laboratory, but in those early years enticing postdocs to work with me was difficult.

Bob Fishman was unprepared for a person like me. Over the previous eight years, he had built the Department of Neurology into one of the premier departments in the country. He had brought me onto its faculty despite my having had only two years of residency training—something he had never done, nor would do again in his twenty-seven years as chairman. Now he was having second thoughts. "What happens if your work on scrapie is wrong?" he kept saying. "You won't want to see patients every day, and once you're tenured, the department will be responsible for your salary, so I'll have to work in the clinic to make enough money to pay it." The department's only dependable source of outside income was professional fees from seeing patients, so Fishman's worries were neither irrational nor unfounded: Several of the senior neurology faculty had lost their grant support, and he could see no reason why I would not follow in their footsteps. In fact, I was more likely to fail, because the others had at least picked reasonable problems—or so it seemed to Fishman.

"Bob, if my work falls apart, I will leave UCSF," I told him. "You're right, I don't want to see patients all day long, every day. There are easier ways to make much more money. The real-estate market in California is booming."

"Where do you get such self-confidence?" he would invariably ask me after such exchanges. "Your mother must have thought you could do no wrong."

My vaunted self-confidence was about to plummet. In July 1979, I entered my sixth year at UCSF as an assistant professor. Most assistant professors were promoted to tenure at the end of their fifth year, but I had not been promoted. Fishman could no longer delay the decision on my tenure; he was faced with promoting or firing me.

Four years earlier, I had been awarded a research career development award from the NIH to defray much of my salary; shortly thereafter, I had returned the funds in order to accept funding from the Howard Hughes Medical Institute. As a Hughes investigator, I received all of my salary from the institute. Holly Smith, the chairman of medicine, had chosen me for a Hughes award because he was convinced that I was a good physician-scientist who would eventually see this scrapie stuff as nothing more than a passing fancy, come to my senses, and work on glutamate in the brain. Now I was told by HHMI that my investigator status would be terminated, since the institute was not funding neuroscience studies any longer. Six months later, Hughes announced a major thrust in support of the neurosciences, which continues today. I learned some years later that Smith was instrumental in HHMI's decision to terminate my award; he wanted to use the Hughes money to support an up-and-coming faculty member in the Medicine Department. Clearly, I had *not* come to my senses; the Hughes Institute would use its money to fund someone with a more reasonable research agenda.

The award's withdrawal created a living hell that lasted more than a year while the verdict on my promotion was in flux. Fishman considered my loss of HHMI support as a major reason to deny me tenure. There were supposed to be some smart people giving advice to the Hughes Institute, and if they had decided to terminate me, then something must be wrong with me. The Hughes support, once so welcome, had proved a Trojan horse. Fishman sent me a letter describing the unfavorable deliberations of the senior faculty.

SAN FRANCISCO: SCHOOL OF MEDICINE
DEPARTMENT OF NEUROLOGY
February 7, 1980
Stanley Prusiner, M.D.
794-M

Dear Stan,

I am writing, upon the recommendation of Dean Reinhardt, to formally summarize our several conversations over recent months regarding your departmental status.

In view of your appointment as an Assistant Professor in Residence in July 1974, your overall performance was extensively reviewed by the senior faculty of the department in recent months. I will summarize these deliberations below. Your efforts have been largely devoted to research and this will be discussed followed by your teaching and service performance.

Research: Your laboratory has been established along two major lines, one an extension of your earlier work in glutamine metabolism, and the other an effort to chemically characterize the scrapie agent. Your expanding bibliography and your ability to obtain extensive financial support for your work was considered very favorably, as was your commitment to develop an innovative research program. However, serious concern was expressed regarding the ability of the department to develop and maintain the kind of laboratory program in scrapie research that you have envisaged. I have assigned to you, on a temporary basis, one third of our research space in the towers. The large assignment of space has been considered excessive in light of the department's overall needs. The question was raised whether our small department could provide extensive support for the broad research program that you demand. These complex issues were not resolved but it was concluded that the reservations of several of the senior faculty concerning the magnitude of your research program should be brought to your attention. You are urged not to recruit more research staff than you can reasonably accommodate.

Teaching: The demands upon your time for teaching of clinical neurology have not been great. You have carried out your assignments responsibly. It was concluded that you have met the needs of medical students much better than the needs of residents. It was recognized that you have spent most of your time in research and have not developed much depth as a clinician.

Service: Few clinical service demands have been placed upon you in light of your support as a Hughes Fellow. Similarly your interest in the patient care service programs has been minimal. Your very limited participation in the major clinical conferences of the department (neuropathology, Child Neurology, Neuro-radiology clinical conferences) has been noted. Your participation in the clinic has been perfunctory. It is suggested that you take a more active interest in the clinical affairs of the Department.

On balance, the senior faculty concurred in my recommendation for a merit increase to Assistant Professor in Residence, Step IV, effective July 1, 1980, primarily because of the continued promise of your research. It is planned to further evaluate the possibility for a tenure level appointment in the Fall of 1980. At present, no such commitment has been intended. As you know, the Hughes Foundation will end its salary support in June 1981. At present, the department does not know whether other funds will be available for your support thereafter. Our current uncertainty regarding long-term funding for your salary must be emphasized. We will be discussing the various possibilities later this year.

I believe the above summarizes our several conversations. Please let me know if you have any questions.

Sincerely yours,

Robert A. Fishman, M.D.

Professor and Chairman

The letter's intent was clear. I had been an assistant professor at the University of California for five and a half years, and a decision about continued employment was imminent. The university's rules were unambiguous: "The tenure review, which generally occurs in the sixth or seventh year, leads to the Chancellor's final decision on whether or not to grant tenure." That means promotion to tenure after six years if your scholarly activities warrant it. If they do not, then you need to find another job. While the system might seem harsh and even cruel, it contains a large element of due process.

I was particularly downcast by Fishman's criticism of my clinical prowess, since I had recently obtained my board certification in neurology. Fishman had made board certification an issue, and his deliberations about my tenure would have been much easier if I had not passed either the written or oral parts of the board examination on my first attempt. I was unwilling to give him an easy excuse for denying me tenure, and I wanted to take these horrific examinations only once. The thought of retaking them, which many neurologists are forced to do, was extremely unpleasant, especially because the hundreds of hours of study did nothing to further my research goals.

Fishman's letter encapsulated the criticisms that dogged me. Many of the negative critiques arose from disbelief, disdain, and ignorance. I had achieved the status of lightning rod. Some of the clinicians in the department looked down on my teaching and clinical service, despite the good evaluations I had received from both students and residents and the time I spent on the outpatient and inpatient services. With Donald Macrae, I ran a general neurology clinic one afternoon a week, and I was an attending neurologist for the inpatient service three months a year. What I didn't do was read the clinical neurology literature religiously; thus I was not conversant with the latest side effects of second- and third-line therapeutics. My only real failing was my lack of seasoning as a clinician. There were some rare diseases I had not seen, largely because of my truncated residency and my lack of interest in outpatient clinical conferences. I couldn't do everything, and I needed to keep my eye on every detail of my experimental studies. I didn't want to neglect the direction of the work during the months when I was the attending neurologist on the inpatient service.

By 1980, I had ten people working with me in the laboratory: three postdocs and seven research assistants. In addition, I had two people working in an office adjacent to the lab, as well as a full-time person washing glassware. All of the work was occurring in just under one thousand square feet of space—it was crowded! We were bursting at the seams, and I knew that we needed more researchers if we hoped to crack the problem. Although Fishman understood that the excitement I had created about solving the mystery of the scrapie agent could bring in more research money, he was afraid that I would consume what little space had been allocated to neurology.

My friend and colleague Curtis Morris, one of UCSF's leading clinical investigators, took my tenure case first to Holly Smith, who offered to give me a tenured appointment in the Medicine Department. While I was grateful to Smith, I did not see myself working in the Medicine Department of a university whose neurologists thought I wasn't good enough despite my board certification. Next, Curtis went to see Julius Krevans, dean of the medical school, who had helped me on numerous occasions to raise private support (figure 12). Krevans is a kind and thoughtful man who must have been unhappy when he found himself being pushed into a corner. In the late spring of 1979, shortly after learning that my HHMI support would terminate, he had asked me to prepare a brief description of my program so he could present it to some people in New York. The New Yorkers were looking for research programs that might push back the frontiers of interesting diseases. I scrambled to prepare a document, and Krevans took it, along with the descriptions of two other UCSF programs, to New York. I heard nothing for a couple of months, and then Krevans told me that the people in New York were interested in my proposal and were planning to come for a site visit in a few weeks. The New York people turned out to be Frederick Seitz, who was the president emeritus of Rockefeller University, Maclyn McCarty, codiscoverer with Oswald Avery and Colin MacLeod of the function of DNA, and James Shannon, former NIH director for thirteen years during its most dramatic growth phase in the 1960s. The money was to come from R. J. Reynolds, the tobacco company.

After our work was presented to the visitors, we adjourned to walk across the street and have lunch. I was stunned when Seitz pulled me aside and said they would give us three years' worth of major support. In those few seconds, he had violated the customary practice of dithering over a decision, especially one involving high-risk research. I was

Figure 12   Some people who saved my research program and preserved my sanity. At perhaps the most crucial moment, Fred Seitz (top left) and Maclyn McCarty (top right) gave me research funds from R. J. Reynolds to further my work. The support and gentle prodding of Curtis Morris (bottom left), Howard Fields, and Ivan Diamond (see figure 5) undoubtedly influenced the bold decision by Julie Krevans (bottom right) to grant me tenure.

thrilled. The Reynolds money was more than a million dollars over the next three years, seven times the research budget HHMI had provided. I gave no thought to the notion that without tenure I would not be able to make use of these generous funds. But this stamp of approval compelled Krevans to overrule Fishman by renewing the tenure process. Fishman then had to prepare the tenure package, and I was later told that he wrote a rather unenthusiastic letter of promotion. Moreover, he did not have a faculty position available at the time my tenure became an issue. When I informed Donald Macrae of my plight, he told Fishman he would retire early to open up a position for me, an extraordinary gesture of support.

Eventually, Fishman seems to have grown comfortable enjoying some of the reflected glory that my success brought to the Neurology Department and UCSF. Fifteen years later, he would give a handsome account of his initial reservations about me in his UCSF oral history interview:

> When he applied to us for residency in 1971, I was a little concerned about taking this smart guy who had been such an unhappy intern and had been in the lab for three years, and I wondered whether he could function well. . . . I remember getting a note from Holly [Smith] who got to know him as his intern in Medicine and he sent me the standard, "He did a satisfactory job" and so forth. Holly and I had talked about him, but then Holly had penned me a note—and it's in Stan's personnel file—"Take him, Bob. He's worth the prickles."[2]

When Fishman asked Mike Bishop, UCSF's renowned virologist, to write a letter in support of my promotion, he demurred. Bishop's superb work on oncogenes with Harold Varmus would later earn them a Nobel Prize in 1989. As I've already noted, Bishop's skeptical view of my work was shared by many of his colleagues at UCSF. The excellence of UCSF's intellectual milieu was greatly enhanced by some of these scientists; it was a milieu which I was at once proud of and at odds with. In his essay "The Planning of Science," Lewis Thomas captures the prevailing view toward my work on scrapie:

> Somehow, the atmosphere has to be set so that a disquieting sense of being wrong is the normal attitude of the investigators. It has to be taken for granted that the only way in is by riding the unencumbered human imagination, with the special rigor required for recognizing that something can be highly improbable, maybe almost impossible, and at the same time true.
>
> Locally, a good way to tell how the work is going is to listen in the corridors. If you hear the word "Impossible!" spoken as an expletive, followed by laughter, you will know that someone's orderly research plan is coming along nicely.

"Impossible" was a word used to describe my research by many people at UCSF (and elsewhere) in those very years when the most unexpected results were emerging.

Thanks to Krevans, by June 1980 my promotion had been approved, and Fishman had apologized for his recalcitrance. In retrospect, I can see that his task was difficult: How could he be supportive, when so many people were so negative about my work? He had neither the training nor the background that would have allowed him to evaluate my work independently and defend it against a formidable legion of doubters. I was greatly relieved; reconstituting my research at another institution would have been extraordinarily difficult. I would have had to acquire animal-care space all over again, recruit a new group of research assistants and postdocs, and build new specialized laboratories for the scrapie studies. I had friends awaiting tenure who were not so fortunate. When it was denied, their professional careers were devastated; they never seemed to hit their stride again in science. The same might well have happened to me.

In the meantime, I continued to be concerned about the amount of lab space available to me. My animal-care problems were under control at Berkeley, but we needed more elbow room at UCSF. The campus was a victim of its own success; there was not enough space to accommodate the needs of many faculty. Some worthy professors had offices the size of a closet, but I had to concentrate on my own problems. I needed space for enough people to produce sufficient data to justify the overall costs of the work. As in all research—and especially in work on scrapie, because of the prolonged incubation times required to assay for infectivity—there is a delicate relationship between funding and productivity: Without sufficient new results, your grant support is endangered, and new results do not appear in a vacuum. Smart people generate new findings by doing clever experiments—and they require adequate space in which to work.

I tried one scheme after another on Fishman, who was responsible for allocating space for his faculty. He must have found my persistent pursuit of lab territory annoying, but I hoped he might also be amused, even impressed, by my dogged determination. I persuaded him to make several visits with me to see Julie Krevans, who invariably declared that if he had the space he would give it to me, but that he had none to give. On one of these occasions, Fishman looked at me as we left Krevans's office and said, "These are issues over which chairmen resign." Without thinking twice, I said, "Well, then, Bob, why don't you threaten to resign so I can get the space I need?" Whenever I recall this exchange, my hubris appalls me. He responded more calmly than he might have: "Stan, I like my job, and I think I'll stay."

I was in turmoil. Where could I turn? How about Frank Sooy, Krevans's superior? This was pretty risky business, but by now I had tenure. Sooy had been apprised of the work five years earlier, when his office had blocked my application to the Kresge Foundation for support to build a biocontainment facility because the university hoped to obtain its own large construction grant from Kresge. After Fishman protested on my behalf, Sooy found money for my facility. Unfortunately, UCSF's application to Kresge was never funded; whether my smaller request would have been funded is unknown.

After I was introduced to Sooy at a luncheon in late October 1980, we became friends. He was an otolaryngologist and had been chairman of the Ear, Nose, and Throat Department before becoming UCSF's chancellor in 1972, a job he had twice turned down before being persuaded to take it by the president of the University of California system. When I told Frank about our work, he immediately grasped its implications for revealing an unprecedented mechanism of disease and said he would be happy to help me in any way he could. This was no time to be bashful. "What I need most is more space," I said. "I don't have sufficient space for my technical staff and my postdocs. There's a critical number of experiments that I need to perform in parallel in order to produce sufficient new data to keep my NIH grants funded. If the generation of new data is too slow, my peers will grow restless and unenthusiastic; their lack of interest in my work will result in my losing my grants."

"Why don't you send me a few papers, and let me see what I can do," Frank said. The next day, I sent him a letter describing some of our more recent findings and enclosed a few reprints of published articles. Despite his offer to help, I knew I could not simply go straight to the top; I had to respect protocol and ask Julie Krevans to speak to his boss. But how was I going to maneuver Krevans into speaking with Sooy? Whenever I needed major help, I would turn to my trusted adviser Curtis Morris, who volunteered to speak to Krevans and in some gentle way prod him to approach Sooy on my behalf. All worked as we had planned. Soon thereafter, Krevans indicated that the Chancellor's Office might be able to help with my space problem.

A few weeks later, the space was found. I was overjoyed—even though it was in an adjoining building, whereas I thought it was important to develop contiguous space so as to minimize the biohazard risks associated with our work. Optimal containment required that the laboratory space be contiguous, with one entrance and one exit. Additionally, contiguous space was the best arrangement for teaching and monitoring good biocontainment practices. I asked William Ganong, the chairman of the Physiology Department, which had some space next to my original space, to trade this real estate for space Sooy and Krevans had assigned to me. My new space adjoined the main block of Physiology, so I was surprised when Ganong was unwilling to make an even trade: He wanted two square feet for every square foot he gave me. When I told him I thought his attitude was uncollegial, he didn't seem to care. I could see that no amount of clever argument would win him over; I was disgusted by his behavior and it was probably written all over my face.

Next, rather nervily, I asked the beleaguered Fishman to move his laboratory to the new space and give me his space, which was next to my original laboratory. He declined, arguing that his current space was superior to what I was offering. My last hope was to approach Howard Fields, who agreed to the move after I volunteered to fund the renovation of his neuroscience laboratory. In the end, all the pushing I had to do was worth it.

7

What's in a Name?

With academic politics behind me, at least for the moment, I could once again focus all my energy on the science. As I drew nearer to unlocking the mysteries of the scrapie agent, however, I would run head on into another kind of politics: the politics of just how you present your discovery to the larger scientific community.

In the meantime, though, while I tracked down the artifacts in the viroid story of Marsh and his collaborators, it dawned on me that we could not afford to concentrate solely on purification. I began to realize that my earlier idea of waiting to characterize the scrapie agent after it was completely purified was naïve: How would we know when purity had been achieved? We needed clues about the composition of the agent. This meant we had to establish, as rapidly as possible, at least one of its essential components. Did it contain DNA, RNA, protein? When we defined one component, we could use it in calculating purity. Once we knew that protein, for example, was a component of the agent, then we could ask, "What percent of the total protein in our preparations is the protein of the scrapie agent?" That is, we could determine whether the scrapie agent in crude brain homogenates was one part in one hundred, one in one thousand, or one in ten thousand. Another reason to probe the structure of the scrapie agent was to ensure that we received appropriate credit for our efforts. After six years of toil, I began to fear that someone else would define the agent's composition based largely on our work.

Once we had enriched our preparations for scrapie infectivity by removing more than 95 percent of the cellular debris, we performed procedures designed to destroy either nucleic acid (DNA or RNA) or protein (figure 13). Enzymes called nucleases degrade DNA and RNA, slicing them up into their constituent bases; some, called DNases, chew up only DNA, while others, called RNases, degrade RNA, and others degrade both. After such a procedure was completed, the sample was injected into four hamsters and the incubation time measured against that caused by a control (untreated) sample. If the time from inoculation to the onset of neurological dysfunction was the same as the control sample, then the nucleases had had no effect. But if the incubation time was prolonged, this indicated that the nucleases had diminished the scrapie agent's infectivity, presumably by destroying its nucleic acid. In contrast to Marsh's results, we found no evidence that a nuclease damaged the scrapie agent.

Whereas nucleases slice up DNA and RNA, other enzymes, called proteases, slice proteins into their constituent amino acids, producing short pieces called peptides. To my delight, the samples exposed to proteases showed a decrease in scrapie infectivity. This meant that the scrapie agent contained an essential protein. We finally had a positive!

We also found that the scrapie agent was more resistant to degradation by proteases than most proteins are. Taking advantage of this relative resistance, we allowed the proteases to chew just long enough to degrade most of the protein but leave the resistant scrapie agent intact. By carefully limiting the time of exposure to proteases, we were able to destroy most of the irrelevant protein but recover the vast majority of the infectious particles, resulting in a large increase in the purity of our samples.[1]

Almost immediately after discovering that protease digestion decreased scrapie infectivity, I began to worry that our findings were due to some subtle artifact. Was the protease contaminated with another enzyme, possibly a special nuclease? But this seemed unlikely after we experimented with several different proteases. Next, we rendered the protease inactive by heating and found that the scrapie agent retained its level of infectivity. Equally important, we showed that increasing the time of protease action gradually decreased the infectivity in hamsters as determined by our incubation-time measurements. And we demonstrated that increasing amounts of protease resulted in progressively lower levels of scrapie agent infectivity. I became more and more convinced that a

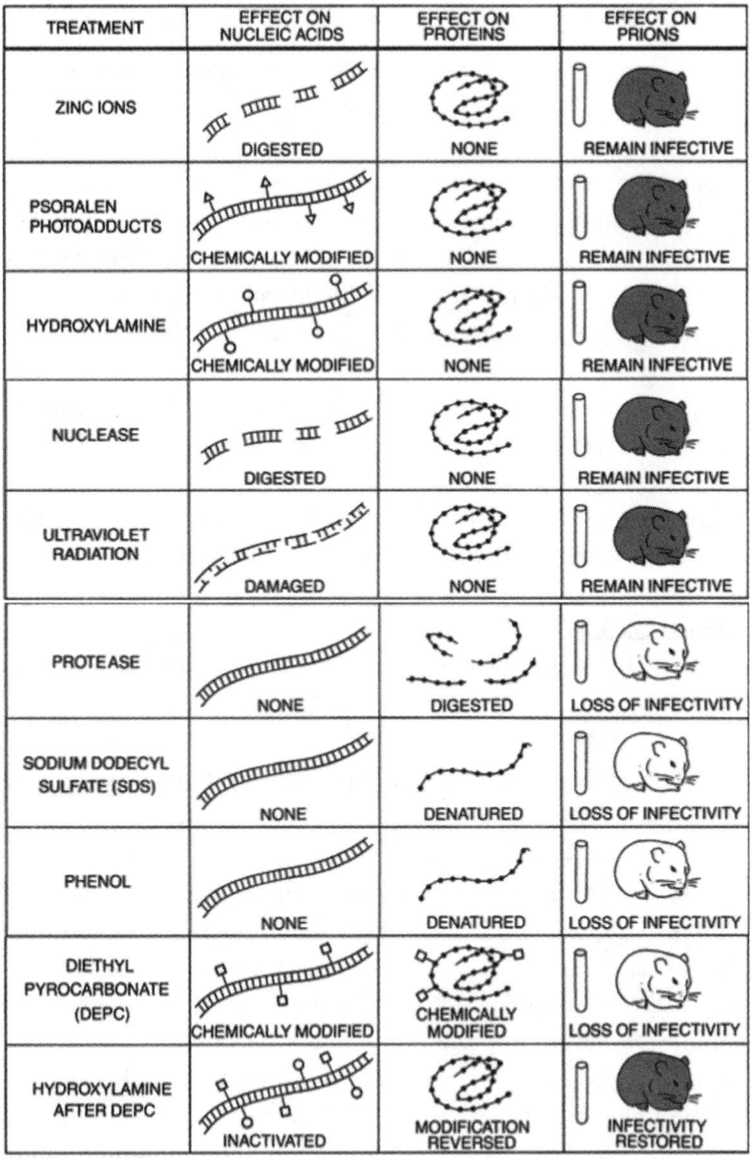

TREATMENT	EFFECT ON NUCLEIC ACIDS	EFFECT ON PROTEINS	EFFECT ON PRIONS
ZINC IONS	DIGESTED	NONE	REMAIN INFECTIVE
PSORALEN PHOTOADDUCTS	CHEMICALLY MODIFIED	NONE	REMAIN INFECTIVE
HYDROXYLAMINE	CHEMICALLY MODIFIED	NONE	REMAIN INFECTIVE
NUCLEASE	DIGESTED	NONE	REMAIN INFECTIVE
ULTRAVIOLET RADIATION	DAMAGED	NONE	REMAIN INFECTIVE
PROTEASE	NONE	DIGESTED	LOSS OF INFECTIVITY
SODIUM DODECYL SULFATE (SDS)	NONE	DENATURED	LOSS OF INFECTIVITY
PHENOL	NONE	DENATURED	LOSS OF INFECTIVITY
DIETHYL PYROCARBONATE (DEPC)	CHEMICALLY MODIFIED	CHEMICALLY MODIFIED	LOSS OF INFECTIVITY
HYDROXYLAMINE AFTER DEPC	INACTIVATED	MODIFICATION REVERSED	INFECTIVITY RESTORED

Figure 13   Preparations enriched for the scrapie infectivity were resistant to inactivation by procedures that selectively altered nucleic acids but susceptible to those that modified proteins. Enzymes called nucleases degrade DNA and RNA, slicing them up into their constituent bases; some, called DNases, chew up only DNA, while others, called RNases, degrade RNA, and yet others degrade both. Six procedures that modified only proteins inactivated the purified scrapie agent. When diethylpyrocarbonate (DEPC) was added to our purified preparations, the scrapie agent was inactivated. DEPC modifies both proteins and nucleic acids, so this experiment, by itself, was unhelpful; however, DEPC could be removed from proteins—but not from nucleic acids—by another chemical, hydroxylamine. Equally important, in the absence of DEPC modification, hydroxylamine reacts with nucleic acids but not with proteins.

protein was essential. Now we could move on to our next step: identifying, by further purification, the scrapie agent protein that was being chewed up by proteases.

As we continued to enrich our preparations, we also determined the amount of protein in them. I knew that our purification was improving whenever the number of infectious units increased and simultaneously the protein content decreased. Future investigations were kaleidoscoping in my head! Once we found the essential protein, I planned to create a one-day test using antibodies that bound only to the scrapie protein, which would enable us to find the gene encoding it and determine whether it was native to the animal or resided inside the infectious agent.

Our findings so far were puzzling. Surely the scrapie agent must also contain an essential genetic component. How could it reproduce inside the animal if it contained no DNA or RNA? I wondered whether a protein coat might be protecting the putative nucleic acid, or whether the presumed DNA or RNA might be specially constructed so as to resist degradation by nucleases. There were undoubtedly still lots of contaminants, any one of which might be preventing the nuclease from cleaving the DNA or RNA.

Concurrent with improving the purification, I decided to begin looking for a nucleic acid using procedures that did *not* involve nucleases. As I began to think more about how to identify the agent's essential components, I recalled my initial reaction to a seemingly cumbersome set of Earl Stadtman's experiments. Some of his work had appeared to be overkill. Stadtman had used eight different experimental approaches to show that a slightly complex ring-shaped chemical group was attached to the enzyme glutamine synthetase. As I reread his seminal paper several times, I wondered why he had been so fanatically thorough. Why would anyone use so many different methods to show that adenylyl chemical groups are attached to the enzyme? And then I had one of those "aha!" moments.

Before Stadtman and his colleagues made their discovery, no one had ever observed the presence of such a chemical group attached to any enzyme—or indeed to any protein—so they wanted to be sure they were interpreting the data correctly. Stadtman didn't want someone else telling him he had made a mistake. In like fashion, I worried that despite the clarity of our experimental results, we might be misinterpreting our data. Thus I set out to find as many different procedures for degrading proteins or nucleic acids as I could. My plan was to identify procedures that selectively modify proteins and use them to damage the proteins in our most highly purified preparations. In parallel, I searched for procedures that selectively modify nucleic acids. In all, I found six procedures that modified only proteins and six that altered only nucleic acids. The procedures modifying only proteins consistently decreased the number of infectious scrapie particles (see figure 13). In contrast, procedures altering nucleic acids had no effect on that number. Since each of the procedures modified either proteins or nucleic acids by different chemical or physical principles, the likelihood of misinterpreting our findings now seemed small.

Interlude: The DEPC and Hydroxylamine Experiments

One of our most convincing experiments was made possible only after we had removed more than 99 percent of the irrelevant proteins. When we added a chemical reagent called diethylpyrocarbonate (DEPC) to our purified preparations (see Figure 13), we found that the scrapie agent was inactivated. DEPC modifies both proteins and nucleic acids, so this experiment, by itself, was unhelpful; however, DEPC can be removed from proteins—but not from nucleic acids—by another chemical, hydroxylamine. Equally important, in the absence of DEPC modification, hydroxylamine reacts with nucleic acids but not with proteins. When we treated our inactive preparations with hydroxylamine, the scrapie infectivity returned. The target of these two chemicals could not be a nucleic acid, it could only be a protein. If it were a nucleic acid, both DEPC and hydroxylamine would have inactivated the scrapie agent. Clearly that was not the case.

Could the scrapie agent consist *only* of a protein? That verged on impossibility. (Shades of Lewis Thomas!) This was beginning to resemble science fiction.

When I described our results to Curtis Morris over lunch one day in the spring of 1981, he said, "Why not summarize all of this in an article for *Science*?" I thought this was a good idea, so I called the editor, Philip Abelson, and told him my story. Our conversation lasted more than an hour and a half. I cautioned Abelson that the article would be controversial—threatening, as it did, the sacred central dogma of biology. "I know how to handle these situations," he replied, so I set to work.

As I began to write the article, I recalled Frank Westheimer's advice. Westheimer, an eminent chemist at Harvard University with a broad knowledge of biology, had visited UCSF for a week in the spring of 1980. Frank was a thin man of medium height with wavy gray hair (figure 14). He spoke with a warm, discerning, and sensitive voice. He was always so honest, straightforward, and enthusiastic. I had known Frank since my days in medical school at Penn, when my aunt Helen Spigel Sacks had introduced us. Frank's wife, Jean, and Helen had begun their friendship in Philadelphia at the age of three. During his visit I showed Frank the results of our centrifugation studies, which fascinated him. When I told him about the experiments involving DEPC and hydroxylamine (see Interlude) and our conclusion that a nucleic acid could not be involved, his face lit up. "You're doing groundbreaking work," he said. "Your data are telling you that the scrapie agent is different from anything that anybody has ever seen before. When you have a better idea about the agent's composition, you'll need to spend some time thinking of a name for it. The name is very important. If you choose a bad name, someone will come along and rename it, and if that happens, your contributions to this discovery may become obscured. But if you give it a good name, it will stick."

The *Science* paper was the perfect place to offer a provisional name for the infectious scrapie particle. I was particularly excited by the idea of extracting the scrapie agent from the world of virology. Its characterization as a "slow virus" or an "unconventional virus" was misleading, since we had no evidence at all for a nucleic-acid core.

But how to choose a name? If the scrapie agent was a brand new kind of pathogen, then this class of infectious agents needed to be denoted by a word that had a fundamental ring to it. The best such words are short, usually comprising one or two syllables. I needed a word like the superb "quark" or the equally good "virus." I began jotting down possibilities, but nothing emerged with the zip I wanted. I pondered consulting someone in UC Berkeley's Linguistics Department. Why had I never taken an advanced course in word derivation? Was there a Greek or a Latin root that might be appropriate? I knew no Greek, and what Latin I remembered proved uninspiring.

The most important words describing the scrapie agent were "protein," "infectious," and "agent." From these, I appropriated a *p*, an *i*, and an *a*, adding an *f* to spell "piaf." While "piaf" lacked that fundamental

Figure 14 Frank Westheimer was a longtime family friend who encouraged me to expend considerable effort creating a good name for the infectious scrapie protein that I discovered.

ring, it was at least brief. I relinquished it when too many people pointed out that it means "sparrow" in French and was the sobriquet of one of the most popular French singers of the twentieth century. When I went back to creating a word, I tossed out "agent" since it was nonspecific, and now I was left with "protein" and "infectious."

These two words obsessed me. One day that spring, I was eating lunch by myself in what was about to become the extinct UCSF Faculty Club in the Millberry Student Union. The view to the north was one of the most beautiful in the world: You could see across the tops of magnificent Monterey pines and cypresses broken by clusters of eucalyptus in Golden Gate Park to the iron-red towers of the Golden Gate Bridge and the headlands of Marin County. As I nibbled on my tuna fish, lettuce, and tomato on mustard-drenched rye, I constructed a two-dimensional matrix with "i-n-f-e-c-t-i-o-u-s" along the horizontal or x-axis and "p-r-o-t-e-i-n" on the vertical or y-axis. This yielded, among some uninteresting gibberish, the word "proin." Not bad, but it had a limp sound to it, whether rhymed with "loin" or pronounced "pro-in." Then I flipped the "o" and the "i" to produce "prion." Eureka! It had a wonderful appearance, perhaps because the last three letters are a fundamental physicochemical term and also occur in "virion." When I pronounced it "PREE-on," it had a satisfying fundamental ring.

I quickly finished lunch and rushed back to my office to look up "prion" in my Webster's unabridged dictionary. To my horror, there it was, a marine petrel found in the Southern Ocean, pronounced "PRY-on," with a long "i." The birds are named for their serrated, or sawtooth, beaks (the Greek word πριων means "saw") (figure 15). Then I thought for a moment. Many words have more than one meaning, and virtually no one outside the ornithological community had ever heard of these petrels. If my assessment of our data proved correct, then my definition and pronunciation of "prion" would take precedence.

Undeterred by a few subantarctic birds with serrated beaks, I proceeded to finish the manuscript. In June, I sent it to *Science* with a letter reminding Abelson that he needed to exercise caution in his choice of reviewers, since the scrapie field was polarized and I had already accumulated a cadre of critics.

Shortly afterward, I delivered a brief talk at the Fifth International Congress of Virology, held in Strasbourg. There I described our findings and explained why these observations argued that scrapie was transmitted by "an infectious proteinaceous particle" (I was not ready to go public with "prion" until the *Science* paper was published). After my talk, the atmosphere was tense. Disbelief filled the room, and the questions were terse—and predictably worded: "How can a virus be only protein?" "How can it be devoid of nucleic acid?"

"Despite the particles causing scrapie often being referred to as 'slow viruses' or 'unconventional viruses,' they are not viruses," I answered. "You just don't know how to isolate the agent causing scrapie." (While I tried to remain calm under fire, I did not always succeed.)

Another query focused on my presumed technical incompetence: "How can anyone who calls himself a scientist not find the nucleic acid core?"

"I have tried to find a nucleic acid using six different approaches," I answered.

"Finding evidence for a nucleic acid is easy. You must have done something wrong!"

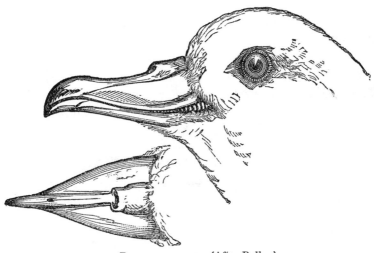

PRION VITTATUS. (After Buller.)

Figure 15 To my horror, there it was: a marine petrel found in the Southern Ocean pronounced PRY-on, with a long I. The birds are named for their serrated, or sawtooth, beaks (the Greek word priwn means "saw").

Still another questioner argued, "Why do you use such odd procedures to purify a virus? You should use a salt gradient composed of cesium chloride."

"I've tried many approaches, including those commonly used to purify viruses, but they simply do not work," I said.

"No one who knows anything about viruses would try isolating a virus by the methods you're employing."

"But I'm not studying a virus. My results clearly support this conclusion."

A few weeks later, David Perlman, the science writer for the *San Francisco Chronicle,* interviewed me about my work. Today, at ninety-three, David Perlman looks the same as he did when I first met him—a man of modest height and medium build. His full head of gray hair crowning his penetrating blue eyes adorned by glasses remains unchanged, as does his warm, kind, inquiring personality. David had a feeling there was a good story hiding somewhere in all these scrapie studies. I explained the scientific findings but continued to keep my new term to myself, promising to send him a copy of the *Science* paper as soon as it was published.

Finally, in September 1981, I received an encouraging letter from Eleanore Butz, a senior editor at *Science*. I needed to respond to the comments of four reviewers. Three reviews were reasonably positive; the fourth reviewer was not and wrote a three-and-a-half-page critique, which began:

> This mixture of fact, unproven hypothesis and unconfirmed personal observations contains some relatively fresh, somewhat brilliant insights into the comparative molecular biology of the scrapie-type agents. . . .
>
> Unfortunately the style is critically flawed and is finally more suitable for a statement of personal opinion—even prejudice—than for a major balanced review in *Science,* e.g.
>
> 1. The style is regrettably chauvinistic, not adequately reporting the work of prior and current investigators.
>
> 2. The main concept, that the spongiform encephalopathy agents are NA-free proteins, depends mainly on the author's own, often unpublished and unconfirmed work.
>
> 3. The term PRION is capitalized as if it is registered as a trademark. It carries unfortunate echos of the author's name (PRusiner IONs).
>
> 4. The early general review of the field is poor, and at no point is the standard term "spongiform encephalopathy" used or explained.
>
> 5. The article has a rather unpleasant, almost sarcastic tone frequently belittling the work of others.

Despite this harangue, *Science* accepted the paper, after I sent back my responses to many of the criticisms: I changed the title of the paper from "Prions Are Novel Microbes: Studies on the Proteinaceous Scrapie Agent" to "Novel Proteinaceous Particles Cause Scrapie" and tried to minimize the use of the word "prion" within the paper. After the paper's acceptance, Butz called me to discuss the manuscript sentence by sentence—an arduous process I suffered without complaint, since I did not want to jeopardize publication.

I was still concerned about the "PRusiner ION" criticism, because it might take hold in the scientific community and detract from what I was trying to accomplish: introducing a label for an entirely novel class of infectious agents that differed from viruses, bacteria, fungi, and parasites. I tried devising another name—one that could not be construed as an act of self-aggrandizement. This wouldn't be easy, because it required discarding "protein" so as to move away from "Prusiner." My heart wasn't in it. I liked "prion," and besides, I had better things to do. I needed to spend my time dreaming up new experiments, deciding on the proper controls, pondering the results, and searching out ambiguities. Meanwhile, I was constantly writing up papers and assembling grant proposals.

Many months went by, and I was still waiting for *Science* to publish the paper. Clearly, someone there was dragging his or her feet. I suspected that Butz, who had tortured me about every word, was worried that the journal's reputation might be marred by my conclusions, and I was afraid to push things, for fear of giving her and the other editors an excuse to retract their acceptance. Then Frank Sooy (figure 16) called David Perlman at the *Chronicle* in hopes that he would cover the Founders' Day celebration coming up in April 1982. Perlman said he would if he were given an advance copy of my *Science* article. When Sooy's office called me, I was only too happy to oblige. Since the embargo policies of scientific journals were poorly developed in those days, I was unconcerned by the prospect of Perlman writing about prions.

Perlman wrote his story and called me on a Thursday afternoon to check some parts of the article while I was attending a Gordon Conference, "Aging of the Brain," in Santa Barbara. The next morning—Friday, February 19, 1982—the *San Francisco Chronicle* carried a front-page story headlined "TINY LIFE FORM FOUND," which continued on the back page with the subheading "WEIRD NEW FORM OF LIFE." On the weekend, the rest of the news media played catch-up. The word "prion" became known worldwide overnight.

Scientists and journalists alike now pushed *Science* to publish my article, wanting to better understand the commotion Perlman had created. (A "weird new form of life"?) Just as I had suspected, Butz, along with a number of other staff members, uneasy about the article's revolutionary implications, had delayed publication, hoping to save *Science* from embarrassment should someone find a nucleic acid buried inside

Figure 16 Frank Sooy (above), then UCSF chancellor, asked to share my manuscript on prions with David Perlman, a reporter for the *San Francisco Chronicle*, before it was published in *Science*.

the scrapie agent. Given the pressure from the media, the editors caved, and my article made its way to the printer. It was published on April 9, 1982, and my coinage had its academic debut, as follows: "In place of such terms as 'unconventional virus' or 'unusual slow virus-like agent,' the term 'prion' (pronounced '*pree*-on') is suggested. Prions are small *pro*teinaceous *in*fectious particles which are resistant to inactivation by most procedures that modify nucleic acids."[2]

By carefully defining prions generically as "particles," I had intended to leave room for another molecule, such as a small nucleic acid. But my competitors would have no part of my waffling. Convinced that I had missed the scrapie agent's nucleic acid, many of them gleefully defined prions as "infectious proteins," paying little attention to the eight years of systematic investigations that underpinned my research. Richard Kimberlin revived the word "virino" to describe one of the possibilities I had suggested for the prion: a small nucleic acid surrounded by protein.[3] Having been pushed into the corner with my "infectious protein," I was forced to fight back, both in scientific papers and during every lecture I gave.

I did not yet know the precise structure of the prion, but I had gone to great lengths to define it as precisely as the available data would permit. I had spent considerable effort describing all possible identifications. My data eliminated the viroid, carbohydrate, and lipid hypotheses. We had shown that the only viable hypotheses on the structure of the scrapie agent were those embracing an essential protein. While I could not eliminate a small nucleic acid buried inside a protein coat, it seemed unlikely that one would be found—but we couldn't prove a negative. The word "prion" provided a challenge to each of my competitors: Find the nucleic acid of the scrapie agent and kill the word as they had redefined it.

Besides suggesting that prions caused scrapie, I also proposed that they cause CJD and kuru, based on data of others arguing that the agents causing all three of these diseases were either identical or very similar. As Bill Hadlow had so presciently written twenty-one years earlier, scrapie and kuru were so alike in their neuropathology that kuru, like scrapie, should be transmissible upon inoculation of brain extracts from dead kuru patients into chimpanzees. Gibbs and Gajdusek had shown the same to be true for CJD. Moreover, working with Raymond Latarjet, Tikvah Alper's former collaborator, they had described the extreme resistance of the kuru, CJD, and scrapie "viruses" to inactivation by gamma radiation.[4]

It is difficult to convey the level of animosity that both the word "prion" and the prion concept engendered. At every turn, I met people who were genuinely irritated by my findings. It seemed that faculty members at many universities were annoyed by their students' fascination with our data, which forced the faculty to take the results seriously and marshal arguments against them. The adjectives used to describe me because of one five-letter word were astonishing. I was called impulsive, presumptuous, reckless, ambitious, aggressive, callous, manipulative, and egotistical. I had thought that people might appreciate the separation our data had effected between the scrapie agent and viruses. And I thought that at least some scientists would enthusiastically embrace a new word that encompassed an entirely new concept in biology and medicine. I was surprised and disappointed by the negative reaction of many of my colleagues. Even some of my friends had difficulty comprehending the prion story as it was evolving. In a commentary that Dick Johnson wrote shortly after I set forth the prion hypothesis, he likened the term "prion" to the naming of a child before he or she is born, noting that his expectant parents had chosen the name "Dorothy" for him while he was still in the womb. When they learned that their baby was a boy, they had to scramble and quickly chose the name "Dick." He suggested that prions were, like himself, named prematurely and maybe they should be called "dorothies" for "*different or other* things . . . with another vowel reversal for euphony."[5] Dick seemed surprised that I did not find either his essay or his acronym funny.

Big new ideas are infrequent in science, especially ones supported by a respectable base of data. After every lecture, I found myself having to argue with members of the audience—sometimes a student, more often a faculty member. "I think it unlikely that scrapie is caused by a virus," I would say, "but I cannot prove you're wrong. That's why I set forth a family of hypotheses. First, I considered the possibility that a prion is composed only of protein. Second, I suggested that it could be a protein and a nucleic acid. And third, I considered whether the nucleic acid might be large or small. Should the nucleic acid be large, then the prion is likely to represent some form of a virus, in which case the term 'prion' should be abandoned."

"Well, I think the virus idea is the only one that's credible!" was the inevitable rejoinder. "It just can't be anything else."

When the discussion became as unproductive as that, I had to stop. I could see no use in continuing to push, since I didn't have the answer. I had only a wealth of data that continued to tell me that the possibility of a virus causing scrapie and CJD was increasingly remote. No matter what others thought, it was critical for me to maintain an objective attitude. Whatever the scrapie agent was, it was. I could not alter its composition. I could only help roll back the shroud of ignorance that had created this conundrum.

Back in 1980, when our new incubation-time assay using hamsters was generating a torrent of data, Theodor Diener, discoverer of the small infectious RNA molecules he named viroids (see chapter 3), offered to collaborate with us (figure 17). Ted's viroids were added to our purified prion preparations, and the mixtures were tested for viroid infectivity in plants at his laboratory in Beltsville, Maryland. We also tested these preparations for prion infectivity in hamsters. More than a year passed before all the experiments were completed. As the data emerged, the answers were always the same: The properties of the viroid were antithetical

Figure 17 Ted Diener discovered viroids, which are small, highly stable infectious RNAs.

to those of the prion. Each time that the viroid was inactivated by a procedure that modified RNA, the infectivity of the prion was preserved. Each time that the prion was inactivated by a procedure that modified protein, the infectivity of the viroid was preserved. Ted's viroids provided an important control. Not only were prions not viruses but they were also not viroids. Our work with Ted, entitled "Viroids and Prions," was published in *Proceedings of the National Academy of Sciences* in September 1982—the first paper in which the word "prion" appeared in the title.[6]

But the results of our collaboration did not put an end to the debate about whether or not scrapie was caused by a virus, perhaps one composed of a small nucleic acid protected by a thick protein coat. I spent many more years searching for the mythical nucleic acid of the prion; if my conclusion that prions were devoid of nucleic acid was to be overturned, I wanted to be the one to do it. One or another of my postdocs would occasionally come to my office with "bad news." Looking gloomy and agitated, they would mumble, "I'm extremely sorry to tell you that I have evidence for a scrapie-specific nucleic acid." I would reply, much to their amazement, that it was far better for us to find the nucleic acid first and report it before one of our competitors did, and I would encourage them to repeat their experiment a couple of times before we decided what else to do. As it happened, none of those leads ever panned out; moreover, many investigators, unable to find the alleged nucleic acid using a variety of approaches, were independently confirming our data.

Why were so many biologists, especially virologists, resistant to the concept of an infectious protein? I asked myself repeatedly. Was their current skepticism a case of unconscious resistance? It had taken biology a long time to disabuse itself of the notion that genes were proteins, an idea that had dominated thinking in biology for more than half a century. Seventy-five years elapsed from the time DNA was first isolated by the Swiss physician Friedrich Miescher until Oswald Avery, Colin MacLeod, and Maclyn McCarty, working at the Rockefeller Institute, showed in 1944 that large fragments of chromosomal DNA were able to transform non-virulent *Pneumococci* bacteria into disease-causing ones.[7] Acceptance of Avery, MacLeod, and McCarty's demonstration that "the transforming principle" was composed of DNA was slow. In fact, it took more than another decade for nucleic acid to be widely accepted as the genetic material of life. This vignette reminds me of Maurice Maeterlinck's dictum: "At every crossway on the road that leads to the future, tradition has placed, against each of us, ten thousand men to guard the past."

8

Lost in the Pacific Fog

O ne of the most satisfying parts of my work was identifying the scrapie agent's essential protein. Around the time of publication of my paper introducing the word "prion," we succeeded in identifying the scrapie protein. With the discovery of what I called the prion protein or PrP, all the tools of modern biology became available. Once we made antibodies, we could perform a limitless number of experiments in several days or at most a few weeks. My research took on an entirely new flavor—no longer were our studies governed by what was possible. Rather, we now had to decide what was most important to pursue from a constantly expanding menu of fascinating choices.

The best strategy for protein identification was to incorporate radioactive atoms as markers into all of the proteins in our purified preparations and then separate the various proteins by mass, using gelatin and an electric field—a process known as gel electrophoresis. After labeling the proteins, we could look for the one that was present in infectious fractions but absent from control fractions prepared from the brains of uninfected hamsters.

But there was a proviso with the gel electrophoresis approach: To use electrophoresis for separating proteins, we had to unfold them—a process known as denaturation, denoting the conversion of a protein from its native, active state to an unfolded, inactive state. One of the most effective procedures for denaturing proteins is to boil them for a few minutes in a mixture of SDS (the detergent sodium dodecyl sulfate) and urea. As the denatured, SDS-coated proteins sieve through the gelatin, they separate according to weight; the smaller ones migrate the fastest, the larger ones more slowly. Unfortunately, both SDS and urea destroyed scrapie infectivity, as measured in the hamsters. Although some proteins can be refolded into an active state, we were unable to recover prion infectivity after gel electrophoresis.

I examined a variety of methods for inserting radioactive atoms into proteins. Was the essential scrapie protein unusual in that it was composed of only a subset of amino-acid building blocks, or did it have all twenty of those that occur in proteins? Each method for inserting radioactive atoms use a different chemistry—each targets a single, specific amino acid. I thought we should use several of these methods, since we had no reason a priori to believe that one technique would work better than another. But there is always a danger in taking too broad an approach; your effort becomes diffuse and you lose focus when the results continue to be negative. While scientists are supposed to relish the unknown, most do not, fearing that their "productivity" will diminish.

I felt I was edging out along the branch of a windswept Monterey pine perched atop a cliff high above the Pacific on a cold, foggy night. I could not see where I was going. Every move had to be made carefully or I might fall into the roaring surf below. If I made the wrong choice of chemical reagents, I could easily fail to insert a radioactive atom and wander aimlessly. As I tried to project my thinking beyond the next set of experiments and consider every imaginable possibility depending on which of several results might emerge, I felt unequal to the task. Sometimes the data from an experiment would be so unanticipated that the preceding speculation was of no value. Yet the best outcome of an experiment is an unexpected result. As this process, which is characteristic of groundbreaking research, repeatedly occurred, my thoughts about the field were continually being revised.

The only sure approach was to use radioactive DEPC, which we had shown (see chapter 7) reversibly inactivates scrapie infectivity. Unfortunately, I did not have the synthetic chemical skills to create it. Until we could figure out how to make radioactive DEPC, I was obliged to use commercially available chemical procedures for radiolabeling proteins. I chose a method that tags the amino groups with radioactive iodine, but I had two major concerns: The amino groups might be inaccessible to the radioactive iodine, and the number of scrapie proteins constituting an infectious unit might be too low to be detectable even if they were labeled. I had no idea whether the scrapie agent was as highly efficient as, say, viruses that attack bacteria, where a single virus particle is sufficient to establish infection. One scrapie protein per infectious unit would represent the lowest possible number—or the number might be ten thousand, like the number of polioviruses needed to infect a mouse.

Moreover, I had no idea about the size of the scrapie protein—or whether, indeed, it came in a number of sizes. It could have been as small as insulin, which comprises 51 amino acids, or larger than the mammoth muscle protein titin, which has more than 30,000 amino acids. The extreme resistance of the scrapie agent to inactivation by heat and chemicals such as formaldehyde argued against a protein complex. Since most proteins are composed of 150 to 300 amino acids, and to keep the number of possibilities manageable, I assumed that there was one scrapie protein of intermediate size. If the pores of our electrophoresis gel were too large, the scrapie protein would travel too fast and migrate out the bottom of the gel. If the pores were too small, the protein would not enter the gel at all and remain as a smear on the top. To address this issue, we ran gels with several different pore sizes. By the spring of 1981, we had occasionally seen a band in the scrapie samples—but not in the controls—of proteins that were around 30,000 daltons.

Interlude: Sizing Up Proteins

The size of proteins is measured in daltons (Da) in honor of John Dalton, who suggested in 1803 that atomic weights be measured relative to that of hydrogen, which was assigned a mass unit of one. In 1961, one dalton was set as one-twelfth of the mass of a carbon atom, which has an atomic weight of 12.011 mass units. Generally, the size of a protein is described as its molecular weight with the units as daltons or thousands of daltons (kDa). Each amino acid has a molecular weight of roughly 100 Da, so a protein of 150 amino acids has a molecular weight of approximately 15,000 Da or 15 kDa.

But our sightings of this labeled band were inconsistent. One explanation was that sometimes there was just enough scrapie protein to detect and other times it was just below the amount needed. In June 1981, I presented our work at a meeting of the UCSF Biochemistry Department at the Asilomar Conference Center in Monterey. One member of the faculty in the Physiology Department, Juan Korenbrot, suggested using a specially configured sucrose gradient for electrophoresis, and that strategy proved to be the key. It increased the purity of our samples, and it could be readily scaled up to produce sufficient pure protein for an array of studies, from measuring infectivity to determining amino-acid sequences.

About that time, David Bolton came to my laboratory as a postdoc; his Ph.D. training had been in virology at the University of California, Davis. David had long, thick, straight, brownish-blond hair that curled inward at the neck, seemingly left over from his days as a rock band musician. On one hand he was enthusiastic about the work, and on the other, he seemed a bit distant and remote at times. While he understood the importance of the studies that I assigned him to do, he seemed to me to have trouble devoting himself completely to his work. I am not sure that I ever understood David sufficiently well to lead him through the journey of discovery that we embarked upon together. Curiously, it was David who brought Thomas Kuhn's book *The Structure of Scientific Revolutions* to my attention as the prion story was beginning to unravel.

When David joined my laboratory, I suggested that he learn to synthesize radioactive DEPC. The synthesis of DEPC proved complicated, so eventually Dave pursued radioactive-iodination studies using commercially available chemicals. Concerned about our earlier inconsistent findings, I asked Dave to prepare analytical electrophoresis gels that would be optimal for detecting proteins of 30 kDa. Dave worked closely with another postdoc, Michael McKinley; using proteases to purify the radio-iodinated scrapie preparations, they repeatedly found a protein of 30 kDa in purified fractions prepared from the brains of hamsters with scrapie— but not in fractions from uninfected control hamsters. Eight months after I had introduced the prion concept in the *Science* article, we had sufficient data for two publications: one in *Science* about the discovery of the protein[1] and the other documenting the purification protocol we had developed.[2]

Mike McKinley had joined my laboratory a few years earlier and was invaluable in advancing some of the work. A tall man with light brown hair parted on the left and combed to minimize his balding scalp, Mike was warm, enthusiastic, and excited about the work. He seemed thrilled to be part of the larger group and was often a great sounding board whenever I was concerned about the progress of our work. I genuinely enjoyed my many interactions with Mike, whose enthusiasm was an important asset for the laboratory.

Having found a readily measurable marker that seemed to track with scrapie infectivity, we needed to determine whether the protein was an artifact or an essential component of the infectious scrapie particle. First, I asked whether it was an adventitious protein—that is, one with properties enough like those of the scrapie agent that it behaved similarly in the isolation procedure we had developed. Second, I entertained the possibility that the protein was produced in response to scrapie infection but was unrelated to the infectious particle. Third was the possibility that the protein was an essential component of the prion—or, indeed, was the prion itself!

We reasoned that if the scrapie protein was a component of the infectious prion particle, then the level of this protein should correlate with the number of infectious units, which we had determined by our incubation-time assay in hamsters.

Interlude: Pinning Down the Scrapie Protein

To undertake such a study, we first had to purify substantial quantities of scrapie prions. Using these highly enriched preparations, we labeled the protein with radioactive iodine and then exposed it to increasing amounts of protease. Low amounts of protease did not alter the levels of scrapie protein as measured by gel electrophoresis. Intermediate amounts of protease decreased the levels of scrapie protein, and high amounts completely chewed up the protein. Next, we inoculated hamsters to determine the levels of prion infectivity in the samples digested with low, intermediate, and high amounts of protease. Low amounts of protease had a minimal effect on both the levels of the scrapie protein and the titer (or number) of infectious prions. Intermediate amounts of protease clearly diminished levels of both the scrapie protein and the titer of prions, while high amounts of protease completely destroyed all the detectable scrapie protein and prion infectivity (see figure 13). This indicated that the unusual resistance of the scrapie protein to degradation by protease was indistinguishable from that exhibited by the infectivity of the prion as measured in hamsters.

Indeed, this turned out to be the case, and I was so enthusiastic about our findings that I wanted to publish them in a high-profile journal. The paper needed to be substantially longer than those typically published in *Science* and *Nature*. I recalled a letter, three years earlier, from the editor of *Cell*, Benjamin Lewin, in which he politely returned two papers and asked me to send him a paper once we had purified the scrapie agent. Now we had not only purified the agent but also identified the protein component. When I called Lewin, he said he was interested, and a couple of weeks later he accepted our manuscript for publication.

No Time for Back Pain

While 1982 had been an extraordinary year, with the introduction of the prion concept and the discovery of the protein, the first half of 1983 was a time to forget. Life had caught up with me. Perhaps I had been working too hard, and I needed to slow down. Again, I was lucky. I developed a herniated disk in my lower back, which was vastly preferable to a heart attack. I suffered from the typical doctor syndrome: It's natural for other people to get sick—after all, they're my patients—but I don't have time for this pain in my back. After about four weeks of intermittent bed rest, I developed a paralysis of the muscles that raised my left foot. This condition is called "foot drop" and is an immediate indication for surgery after a magnetic resonance imaging scan of the spine to identify the cause. In those days, magnetic resonance imaging scanners were relatively new, though, and I chose not to travel ten miles to have an experimental scan performed. Instead, Bob Fishman, who was caring for me, decided to have a CT scan done, which showed no lesion. Bob hoped that the foot drop would slowly disappear, and I was in no position to argue—I was prone, in fact, and he was finally in charge. After three more weeks of bed rest, I decided that dragging my left foot around for the rest of my life was a bad idea. I mounted the courage to tell Fishman that I was going to ask my neurosurgical colleague and friend Phil Weinstein to consider operating. Removing whatever was causing my paralyzed foot seemed rational. My only worry was the extent to which I would be able to recover movement in it.

Phil admitted me to the hospital the next day, and a myelogram showed that a fragment of one of my lumbar disks had broken loose and was encroaching on a nerve root. The diagnosis was clear: The cause of my foot drop was a disk fragment that needed to be removed. While recovering from the surgery, which turned out to be successful, I put the finishing touches on the *Cell* paper. Dave Bolton and Mike McKinley came to my hospital room and said we needed a name for the scrapie protein. They suggested "Scp 27–30," with "Scp" derived from *Scrapie protein* and "27–30" because the protein showed up in the gel electrophoresis as a diffuse band with a molecular size of 27 to 30 kDa.

I, too, had been pondering the issue of a name for our essential protein, but I was not always thinking clearly, because of the pain medications. After leaving the hospital, I decided we could come up with a better one. I called Dave and Mike. "I don't want to restrict the name of the protein to scrapie by using the abbreviation Scp," I said, "since we and others will surely find the same protein in other diseases, like kuru and Creutzfeldt-Jakob. We need a more general name. Let's use 'prion protein,' and we'll abbreviate it to PrP. In our *Cell* paper, why don't we call it PrP 27–30?" I thought this might be a good compromise that would create a more widely applicable terminology.

There was silence for a moment and a perceptible coolness in David's voice. I wondered whether he didn't want the word "prion" associated with the protein he and Mike were working on. Perhaps, like so many others, he was apprehensive that the prion concept might not hold up; after all, he had been trained as a virologist.

Despite the clarity of our data, alternative interpretations of our findings abounded, and we tried to acknowledge all of them in the discussion section of the manuscript. Generally, scientists consider as many alternatives as they can contemplate and then try to eliminate various possibilities as more data accumulate. So it was with the protein PrP 27–30. Was PrP 27–30 a component of the infectious prion? Might it be the sole component? Was PrP 27–30 a product of prion infection and unrelated to the infectious prion particle? I thought the last possibility so remote as not to be worth mentioning, but both Bolton and the referees at *Cell* thought this possibility should be mentioned, and so it was.[3]

One way to pin down the significance of PrP 27–30 was to look for it in other diseases, like kuru and CJD, but this was difficult, because we would have had to isolate the protein from human patients or infected chimpanzees before attaching radioactive iodine atoms. Although our intermediate biocontainment facility was superb for scrapie studies, I was worried that it was inadequate for studies on a human infectious agent.

We also lacked antibodies that reacted with PrP 27–30. A technique called Western blotting could be used to detect PrP 27–30 from almost any source once we had antibodies, which are so specific that we would be able to find PrP 27–30 in brain homogenates or even in thinly sliced brain sections under a microscope. To produce these antibodies, we injected rabbits with our most purified fractions, but repeated booster injections produced no immune response. The rabbits refused to make antibodies! Although I had hoped they would, I was not surprised; the literature was replete with reports on the lack of an immune response to infection with the scrapie agent. Sheep, goats, rats, mice, and hamsters all developed the scrapie disease without provoking any reaction in their immune systems. The same is true for kuru and CJD in humans. Not only were there no antibody deposits in the brains of animals with scrapie, but there also was no evidence of a cellular immune response involving lymphocytes. Nor did the interferon system, which is activated by viruses, respond to scrapie prions.[4]

One of the features shared by scrapie and CJD that intrigued me was this quiescence of the host defense systems. My CJD patient had died without ever developing a fever. She never showed an elevated white-blood-cell count, either in her blood or her cerebrospinal fluid. At autopsy, her brain showed no evidence of antibody deposits or white-blood-cell infiltration. She died of a "prion infection," but her defense systems were silent. Had she been infected by viruses, bacteria, fungi, or parasites, her defense systems would have been activated, she would have developed a fever, and she would have had hundreds, or thousands, of lymphocytes in her cerebrospinal fluid. Since proteins are generally the best of all molecules at stimulating an immune response, our finding that scrapie infectivity requires a medium-size protein was unexpected and somewhat puzzling, though our failure to produce antibodies in the rabbits was consistent with the stealthy nature of the scrapie agent. I was back in the Pacific fog once again!

George Balanchine

In many respects, the case of my CJD patient paralleled that of George Balanchine (figure 18), the undisputed master of the ballet, who died on April 30, 1983, while we were focused on producing antibodies and defining the properties of PrP 27–30. In contrast to my patient, who developed a rapidly progressive dementia and died three months after I initially examined her, Balanchine's illness had been slow and mysterious. He began complaining of unsteadiness in September 1978 but was still able, at age seventy-four, to perform dance movements requiring a high level of coordination. After triple-bypass heart surgery in March 1979, his angina disappeared but his difficulties with balance persisted. Over the next three years, Balanchine's trouble executing pirouettes and other dance maneuvers increased, but he was in his middle seventies, so some decline in his physical prowess was not unexpected. His brain was challenged daily as he imagined a dance routine, described the choreography to his dancers, and showed them the complex movements he expected them to perform. In the spring of 1982, his balance began to worsen. Several CT brain scans and many neurological consultations in New York City failed to produce a diagnosis. His balance problem was described as "truncal ataxia," the very term used to describe the balance difficulties seen in kuru patients late in the progression of their disease.

In October, Balanchine fell in Washington, D.C., while on tour with the New York City Ballet. He was hospitalized there, but no limb ataxia was found in his arms and legs, even though his truncal ataxia was profound. While his difficulties with balance were similar to those of the Fore with kuru, the coordinated movements of his arms and legs were unaffected. In contrast to Balanchine, all fifteen kuru patients I saw in New Guinea had showed marked difficulties in coordination. Like the New York neurologists, his Washington doctors were baffled.

After returning to New York, Balanchine fell again and was admitted to Roosevelt Hospital on November 4. The consulting neurologist, Sidney Bender, found that Balanchine knew that he was in Roosevelt Hospital and that Ronald Reagan was the president of the United States, but not what year it was. Consistent with a dementia were prominent snout, rooting, and grasping reflexes seen in kuru patients late in their illnesses and in most other demented people. Balanchine's truncal ataxia was profound, but his limbs were still relatively unaffected; moreover, the muscle strength and tone in all his extremities was remarkably good for a seventy-eight-year-old. Over the next six months, he remained hospitalized, and his health steadily worsened: His ataxia became severe, his memory for

Figure 18   Choreographer and ballet master George Balanchine began to complain of difficulties with his balance in 1978, nearly five years before he died of an elusive, progressive neurological disease on April 30, 1983.

recent events vanished, and he was often confused and combative. By February 1983, he was unable to speak, though he still had moments of lucidity.

Since the disease that killed him was undiagnosed at the time of death, an autopsy was done. A diagnosis emerged only after the brain was sent to Columbia University and the neuropathologist Philip Duffy prepared sections for microscopic examination. Duffy found severe spongiform degeneration in the cerebellum, along with numerous "amyloid" plaques (also called kuru plaques). In addition, mild sponginess was seen throughout the cerebral hemispheres. The diagnosis seemed clear: This was Creutzfeldt-Jakob disease—more specifically, a rare form called ataxic CJD.[5]

The amyloid plaques found in Balanchine's brain, while common in kuru victims, were found in only about one in ten people who died of CJD. When pathologists used Congo red dye to stain thinly sliced sections of brain, these plaques, as well as those seen in all cases of Alzheimer's disease, were classified as amyloid. While the presence of such plaques was helpful with the diagnosis of these dementing diseases, pathologists thought that the plaques were filled with "junk," composed of nonspecific lipids, sugars, and proteins, which tend to accumulate with age. Was this supposition true? Exactly what were these amyloid plaques made of? I would soon delve into this line of inquiry.

9

The Amyloid Story

O ur discovery that prions form amyloid fibrils that coalesce into plaques would foreshadow by several decades a series of discoveries eventually showing that prions cause a wide array of neurodegenerative diseases. The idea that prions could form amyloid unleashed another maelstrom of controversy. For me, the amyloid discovery was a remarkable finding that created a kaleidoscope full of unimaginable, future, unanticipated findings.

The combined funds from the NIH and R. J. Reynolds had provided adequate support for our efforts in the late 1970s, but more money was needed if we were to scale up our efforts, which I had been thinking about doing long before we discovered PrP 27–30. My work on glutamine enzymes at the NIH had taught me the value of large-scale preparations. Unless you have adequate quantities of purified protein, your experiments become too conservative: You tend to choose those that are likely to "work."

At the weekly laboratory meeting where we reviewed our progress, I asked the members of my group whether they could think of anyone to help us raise the needed funds. Paul Bendheim, an obliging young UCSF neurologist, said, "You need to talk to my cousin's boss." His cousin's boss, a lawyer with numerous connections in the information-technology business, suggested that the Sherman Fairchild Foundation might be a source of funds and that I should meet with Lester Hogan, the former president of Fairchild Industries. This I did, and Hogan, a physicist, be-

Figure 19 Walter Burke (right), the president of the Fairchild Foundation, became an enthusiastic benefactor of our work. Photograph taken at Dartmouth Commencement Ceremony, May 1998.

came enthusiastic when he learned that Frederick Seitz had been supportive of my work; he had used Seitz's textbook, *The Modern Theory of Solids*, when he was a graduate student at Harvard in the late 1940s. The next day, he telephoned Walter Burke (figure 19), the president of the Fairchild Foundation, and a few days later Burke called to ask me to meet with him in New York City in order to discuss my needs.

In October 1982, I flew to New York to explain the scrapie project to Burke, emphasizing that our studies would shed light on the etiology of kuru and CJD and perhaps lead to a cure. Walter Burke was a man of medium build and height with straight brown hair parted on the left. Walter gave forth a thoughtful, gentle gaze but was always trying to envision the next ten steps of any process. Behind his warm blue eyes was an extremely intelligent man with wide interests, ranging from the great painters to the exploration of the cosmos. His penetrating questions and his ability to grasp the essence of what you are trying to say may reflect his many years of practicing law. Before we met, he had already discussed our efforts with Hogan, and now he listened carefully as I described my vision. "You can have any seat in the ship," he said when I'd finished—an elliptical remark that I interpreted favorably. The Burkes and I later became good friends. Whenever I compliment Walter on his courage in supporting our work while many of my colleagues were still skeptical, he reminds me that I told him the solution to the scrapie problem was straightforward—something of an exaggeration at the time. In any event, the foundation has contributed many millions to our work over the past three decades.

With the Fairchild Foundation money, which began coming through in December, we could now expand both our biocontainment laboratory for large-scale purification and our animal colony to accommodate hundreds of additional hamsters. To this end, I visited the Lakeview Hamster Colony, in Newfield, New Jersey, where most of the Syrian hamsters destined for pet stores or research laboratories in the United States are raised. There I learned that the largest litters occur with the third and fourth pregnancies and that it would thus be substantially cheaper to buy females well into their third pregnancies than weanlings. So we set up a delivery room at UC Berkeley's superb animal care facility, where I had been housing our hamsters since 1980. We occupied eight large animal rooms on the third floor, seven filled with hamsters and one with mice.

A Huge Leap Forward

By 1980, using the electron microscope, Mike McKinley and I had begun seeing somewhat irregular cylindrical structures with slightly different shapes and a wide variety of lengths in our prion preparations. Neither of us appreciated the significance of these tiny rods, which also turned up—but in far lesser numbers—in control fractions prepared from uninfected hamster brains. Equally perplexing, we could not understand why the rods were in some sucrose-gradient fractions enriched for scrapie infectivity and not others.[1] That was the state of our knowledge at the end of 1982, when the Berkeley biophysicist Robley Williams, who had so kindly communicated my first paper to the *Proceedings of the National Academy of Sciences* five years earlier, took an increased interest in our studies.

Williams, an authority on the structure of viruses, had invented a technique called rotary shadowing for visualizing viruses in the electron microscope. It was based on his earlier work as an astronomer in which he manufactured mirrors for telescopes. Samples subjected to rotary shadowing were spread on a support film, which was rotated during the shadowing process. A heavy metal—typically, platinum, chromium, or uranium—would be deposited on the supporting film, creating an uneven surface that reflected the architecture of the virus. These rotary shadowing studies confirmed our impression of the shape and size of the rods.

It was Robley who described the elongated particles as "rods." Although their size—about 10–20 nanometers in diameter and 100–200 nm in length—placed them well within the range of viruses, Robley didn't think they were viruses, because of their varying size and shape. He had a wry sense of humor and more than once said to me, "When you've seen one tobacco mosaic virus, you've seen them all," TMV being the standard virus used in laboratories to teach students electron microscopy. He was so right! When we mixed TMV into our preparations, we could easily distinguish the regular, uniform TMV particles from the rods (figure 20).

Robley had come to UC Berkeley in 1950 to work with Wendell Stanley, who in the late 1930s had succeeded in crystallizing pure tobacco mosaic virus. The TMV behaved like a protein, in that it could be crystallized. Moreover, chemical analyses of crystalline TMV revealed only protein; its RNA genome, which accounts for about 5 percent of the mass of the virus, was buried inside and went undetected. Thus Stanley concluded—incorrectly—that highly purified TMV preparations were composed only of protein, and he received the Nobel Prize in 1946 for this work. I was never able to get Robley to speak about Stanley's gaffe. Whenever I raised the issue, he scowled and changed the subject.

The rods puzzled me. I couldn't reconcile these large rods with Tikvah Alper's data arguing that the infectious scrapie particle was much smaller. Robley was equally baffled but suggested that Alper's findings might explain why fractions from the bottom of the sucrose gradient had rods, whereas fractions at the top did not. At the bottom of the gradient, many small scrapie prions could have assembled into rods, whereas at the top they remained unaggregated. The high level of scrapie infectivity at the top of the gradient suggested that the prion, when dispersed, was too small to visualize with the electron microscope.

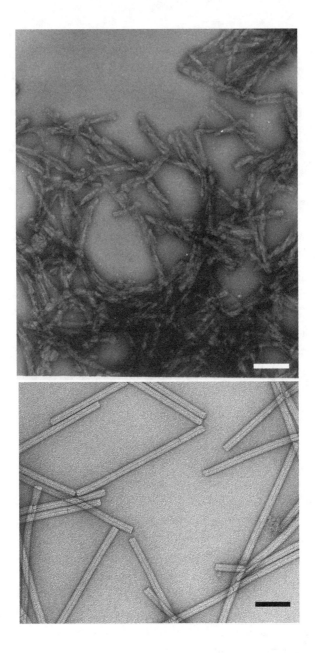

Figure 20 The elongated particles found in some of our purified fractions are called "prion rods" (top). The regular structures of the tobacco mosaic viruses (bottom) contrast with those of the prion rods.

"Are you saying that the rods might be aggregates of prions?"

"That's a possibility," he said. "All we really know is that we can see them. We have to do a hell of a lot better than that."

There were two possible explanations for the rods, which increased in number as our preparations became more enriched for infectivity. They were either contaminants or ordered aggregates of proteinaceous particles—the infectious prions. (Later we learned that our enrichment procedures favored aggregation of the scrapie particles, explaining why the number of rods rose as the purity of preparations increased.) Around this time, in a book entitled *The Electron Microscopy of Proteins,* I came across a chapter on the fibrous protein aggregates known as amyloids.[2] It struck me that the pictures of amyloid fibrils were indistinguishable from the rods in our preparations. They struck Mike and Robley that way too. "Do you know an amyloid expert?" I asked Robley. "No, but I think that you need to find one," he replied.

For almost a century, physicians interested in diseases of the nervous system had been examining brain sections from older people. They often observed small round areas that absorbed stain and seemed to be devoid of neurons. These areas were called amyloid plaques. The most common brain disorder in which amyloid plaques were found was Alzheimer's disease. When Congo red dye was used to stain brain sections from victims of Alzheimer's or kuru, it produced a deep red color that could be seen in the light microscope.[3] When filters that polarize the bright light from the microscope lamp were used, the Congo red bound to the amyloid fibers displayed a characteristic green-gold appearance. That is, when the specimen was rotated 45° in polarized light, the green regions become gold and the gold regions become green. These changes in color that occur every 45° are referred to as birefringence.

For more than two decades, it was known that a minority of CJD patients and a majority of kuru patients had amyloid plaques in their brains, as was the case with George Balanchine. In addition, some scrapie-infected sheep had amyloid plaques, and the 87V strain of the scrapie agent isolated by Alan Dickinson, the irascible director of the University of Edinburgh's Neuropathogenesis Unit, produced large numbers of plaques when injected into the brains of a strain of mice he named VM/Dk.

An Extraordinary Story Unravels

I could readily see an extraordinary story unraveling: Plaques in the brain had been considered collections of junk for almost a century. Did the amyloid plaques of kuru, CJD, and scrapie contain infectious prions? If the answer was affirmative, then this finding would argue that either the plaques themselves or the protein within them caused the disease. But there was a big caveat in this new slant on degenerative neurological diseases: Although Alzheimer's resembled kuru and CJD in many respects, it did not appear to be infectious. Attempts to transmit it to apes and monkeys had been unsuccessful.

I learned that George Glenner, an expert on amyloids at the NIH, was on sabbatical at UC San Diego, and I decided to ask him whether Mike McKinley and I could bring a purified sample of our hamster prions for him to evaluate. To Glenner, my idea that the scrapie agent might be amyloid seemed crazy, but he agreed to look at our preparations. When he examined our samples stained with Congo red, his eyes widened. He was seeing the green-gold birefringence in the polarization microscope. The rods in our preparations, which were highly enriched for prion infectivity, were indeed amyloid.

It all fit together so beautifully. We had found that an amyloid could be infectious, and that it caused scrapie and probably CJD. On the flight home, Mike and I were numb. The prion story might just be opening up the study of all degenerative diseases of the brain. For more than fifty years, these disorders had occupied a biological wastebasket of diseases of unknown causes—including Alzheimer's, Pick's, Huntington's, Lou Gehrig's and Parkinson's diseases. If amyloid plaques consisted of prions, then the proteinaceous deposits in other degenerative brain diseases were likely to be the causes of those diseases.

Now I had to prepare a manuscript. The next day, I called Ben Lewin at *Cell* and told him what we had found. He encouraged me to write the paper quickly and send it to him for consideration. I told George Glenner that I was happy to have him as a coauthor. I also invited Robley to be a coauthor, but he declined: "Stan, this is your work, not mine. You discovered the amyloid connection, not me."

When our paper was published in *Cell* in December 1983, there was considerable media interest.[4] I described our conclusions to the press, telling the *New York Times* that "we never would have dreamed that amyloid and prions were the same. The implications of the findings may be enormous."[5] But Glenner, the most eminent of my six coauthors, denied that it was an important result: "I have the greatest respect for Stanley, but he wanted to get in the press fast," he later told *Discover.* "I do not think it should have been released at all."[6] The media were perplexed: On December 7, 1983, the *MacNeil/Lehrer NewsHour* reported:

> Depending on whom you listen to, there may or may not have been a breakthrough in the search for the cause of Alzheimer's disease, a type of senility that affects two million Americans and is the fourth leading cause of death in the country. Up until now, doctors have not been able to find either the cause or the cure for the disease, but yesterday a team of researchers from the University of California reported that a tiny infectious agent, called a prion, which acts like a virus, may be identical to a substance found in the brains of Alzheimer's victims. Dr. Stanley Prusiner, of the University of California at San Francisco, says this finding may mean the prion actually is the cause of Alzheimer's and other neurodegenerative brain disorders. But one of his coauthors, Dr. George Glenner, of the University's San Diego branch, told us that Dr. Prusiner is jumping to conclusions and that there's no solid evidence to support the theory that the prion may cause Alzheimer's. Dr. Glenner says the research amounted to a very small finding of the similarities between the two substances, and added that it's "mindboggling to draw any conclusions from that." Dr. Prusiner, meanwhile, called it "an astounding finding with enormous implications." Both men, remember, are talking about the same report that they worked on and signed their names to. The report appears in the current issue of the scientific journal *Cell.*

I was furious. Glenner had spent fifteen minutes helping us, and for that I had made him a coauthor. Only later did I learn of speculation that Glenner was unenthusiastic because he was focused on isolating the proteins of the vascular amyloid found in the brains of deceased Alzheimer's patients—a paper that was submitted for publication four months later.[7] No matter how passionate Glenner may have been about his own work, he did not have a means of showing the importance of his discovery. In contrast to our prion rods, which caused neurologic dysfunction in hamsters, Glenner's amyloid protein had no readily measurable biological activity. It turned out to be a peptide of forty amino acids and was subsequently named amyloid beta (Aβ). Aβ was later shown to be pivotal in understanding the pathogenesis of Alzheimer's.

In any event, that prions form amyloid was an unanticipated finding, with remarkable consequences for deciphering the pathogenesis of all neurodegenerative diseases. As a medical student, and later as a neurology resident, I had been taught that amyloid plaques were ancillary, inert junk that accumulated in some central nervous system diseases and also in many healthy older people. Our work had changed that notion forever.[8]

Faced with an ambiguous response from my scientific colleagues, we needed to demonstrate unequivocally the presence of a specific protein—PrP 27–30—within amyloid plaques found in scrapie, kuru, and CJD. The best approach seemed to be to determine this by using antibodies specific to PrP 27–30, but we still had no such antibodies. I encouraged a new postdoc, Ronald A. Barry, who had recently finished his Ph.D. in immunology, to join Paul Bendheim's effort to produce antibodies in rabbits, something we had thus far been unable to do (figure 21).

In January 1984, I flew to Edinburgh to ask Alan Dickinson to collaborate with us. I wanted to purify the amyloid plaques from his VM/Dk mouse brains and determine whether they contained PrP 27–30. Dickinson had asked me to give a seminar on our investigations to his unit. Unexpectedly, he interrupted my lecture at what seemed to be thirty-second intervals, with objections to one or another of my studies' conclusions; clearly, our data did not support his view of what had to be true. After the seminar, and in spite of his annoying behavior, I proposed the collaboration, but he demurred. Nor would he give me access to his VM/Dk mice, since he feared we would be unable to maintain the inbred strain; moreover, he saw no reason to do so, doubting that we had actually found amyloid. Our discovery could not possibly be correct; it had to be an artifact, a stupid mistake.

I had asked Dickinson several times in the past to send us some of his VM/Dk mice for a study of possible genetic factors in scrapie agent incubation times, since the incubation period for the VM/Dk strain was atypical. However, he was reluctant to part with them until he had bred

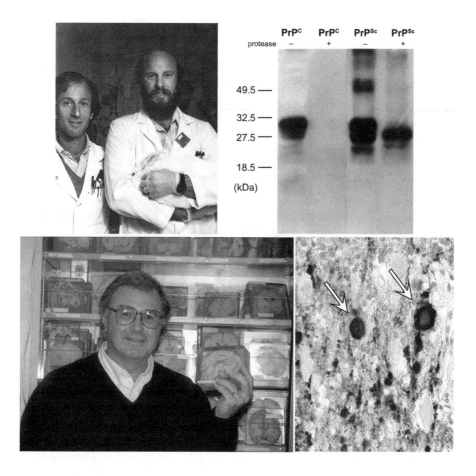

Figure 21   Antibodies raised in rabbits by Paul Bendheim (top left), Ron Barry (top left, holding rabbit), and Dan Stites bound only to PrP (top right) among the numerous proteins found in the brain. Working with me, Barry discovered a new protein, which we called PrP^C, for "cellular PrP." The superscript distinguished it from a protein of the same size found in scrapie-infected brains, which we designated PrP^{Sc}, for "scrapie PrP." Steve DeArmond (bottom left) found that the amyloid plaques in scrapie prion–infected hamster brains were intensely stained with these antibodies. A quarter of a century later, sections of fixed brain tissue from George Balanchine (bottom right) were stained with antibodies indicating PrP amyloid plaques (arrows). Henryk Wisniewski, Heino Diringer, Carleton Gajdusek (see figure 9), Alan Dickinson (see figure 8), and many others championed the idea that amyloid composed of PrP 27–30 was a consequence of the infectious process created by the elusive scrapie virus.

them with a unique coat color. He was concerned that, absent the unique coat, they would get mixed up with other strains. If that were to happen, and if my results differed from his, all hell would break loose. At this point, I was tempted to ask him to make purple mice for us, but I thought better of it. That evening I had dinner with several senior scientists from his unit in a pub adjoining Castle Rock, at the base of the thousand-year-old Edinburgh Castle. Richard Kimberlin, Dickinson's deputy director, apologized for his boss's behavior: "It's just that he's upset by your results," he said. "We all believe that the scrapie agent has a nucleic acid. For one thing, you can't explain the existence of different strains of the agent without it." This was a perfectly reasonable objection, and a decade passed before I was able to answer it (see chapter 15).

While I was in Edinburgh, Ron Barry and Paul Bendheim succeeded in producing the antibodies we needed, in a rabbit that Ron immunized with large amounts of denatured PrP 27–30.[9] They had managed to emulsify the protein with an oily substance called Freund's adjuvant, which is commonly used to stimulate antibody production. Within two months, the PrP antibodies would render my proposal to Dickinson obsolete.

My UCSF colleague Stephen DeArmond, using our PrP antibodies, decided to devote his experimental neuropathology studies to investigations of several potential varieties of prion disease (see figure 21). Steve had been studying a protein called GFAP (glial fibrillary acidic protein), which accumulates in the brain in response to injury. His background in the microscopic anatomy of the brain, his ability to visualize the anatomical pathways in the nervous system, and his extensive experience in staining brain sections with antibodies ideally suited him to exploit our work. He found that the amyloid plaques in scrapie-prion-infected hamster brains were intensely stained with the rabbit antibodies to PrP 27–30.[10] Thus, in one experiment, Steve established what I had hypothesized nine months earlier: Amyloid plaques in scrapie-infected brains contain the prion protein.

Next, Steve showed that the amyloid plaques in CJD and GSS (Gerstmann-Sträussler-Scheinker syndrome, another human prion disease) contained PrP. To be sure that Philip Duffy's diagnosis of George Balanchine's CJD—made nine months before we had antibodies to PrP 27–30—was accurate, we retrieved paraffin blocks containing small pieces of Balanchine's brain from the Pathology Department at Columbia University. Using the antibodies, Steve found that the amyloid plaques in Balanchine's brain stained intensely, indicating that Duffy's diagnosis was correct (see figure 21). Steve also demonstrated that PrP antibodies did *not* stain the plaques of Alzheimer's disease, a result in accord with the work of Glenner and others, and a clear indication that PrP was not simply a nonspecific response to brain injury.[11] In Alzheimer's, it is the Aβ peptide discovered by Glenner that forms fibrils that coalesce into amyloid plaques.

Steve's immunostaining studies were the crown jewels glittering atop our initial finding that Congo red dye binds to purified prion rods and exhibits green-gold birefringence. The PrP 27–30 antibodies allowed us to determine the specificity of PrP deposits in the human brain. Although the amyloid plaques of CJD and Alzheimer's were indistinguishable by Congo red and other standard histological techniques, the immunostaining with our antibodies told us that the plaques were specific to each disease. When Steve—and, later, others—used PrP antibodies on brain sections from victims of several other neurological diseases, there was no staining. The extreme specificity of the PrP deposits demonstrated not just that PrP accumulation was not a nonspecific response to brain injury but also that PrP 27–30 was an essential component of the infectious prion.

The Opposition

Not all of my fellow investigators would agree. Some years earlier, at the New York State Institute for Basic Research in Developmental Disabilities, on Staten Island, the neuropathologist Henryk Wisniewski and his colleagues, using the electron microscope, had found pairs of long, twisted fibrils in detergent extracts of brain homogenates from scrapie-infected mice—but not in uninfected controls. By 1981, Wisniewski and his talented research assistant Patricia Merz had become convinced that their scrapie-associated fibrils—or SAFs, as they were called—were different from amyloid fibrils but somehow related to scrapie. Over the next two years, they would compare SAFs to many different amyloids and publish a detailed study describing the features of SAFs that distinguished them from amyloid.[12] However, because SAFs had been studied only in crude brain fractions, I had no idea how to relate them to our work.

After we published the two 1982 papers describing our discovery of PrP 27–30, Heino Diringer, working at Berlin's Robert Koch Institute, quickly confirmed our findings. For this I was grateful, since no scientific report can be considered authentic until it is independently verified. But Diringer looked at the situation quite differently than I did. He and his coauthor, Richard Kimberlin, concluded that the SAFs were composed of PrP 27–30, and they published their postulate about the same time we published our findings that purified prions were indistinguishable from amyloid.[13] I argued that SAFs had repeatedly been shown to differ from amyloid, so how could they be equivalent to our rods? Wisniewski had reported that SAFs in the electron microscope differed from amyloid and did not stain with Congo red dye. On more than one occasion, I asked Diringer and Kimberlin how they could equate SAFs to the rod-shaped particles in purified prion preparations. The rods in our preparations did not exhibit the distinct twists that defined SAFs. As Wisniewski had shown, these ultrastructural features of SAF were what distinguished them from amyloid. Their response ("But SAFs are composed of PrP 27–30") was frustrating. While I agreed that their purified preparations contained PrP 27–30, I argued that the fibrils in their preparations were not SAFs.

What Diringer and Kimberlin had done was to redefine Wisniewski's SAFs, and in doing so, whether they intended to or not, they had managed to blur the record of discovery. True, SAFs were described two years before our work on amyloid was published, but Wisniewski and his colleagues reported that SAFs were distinct from amyloid and later contended that SAFs were filamentous animal viruses. Moreover, Wisniewski and his coworkers had not purified SAFs, so their composition was unknown. My group had made an important discovery, and it seemed to me that Diringer and Kimberlin were intent on altering the published record.

To Wisniewski's credit, he initially stood behind his contention that SAFs were not amyloid. Later, when it became clear that PrP amyloid was composed of infectious prions and responsible for a number of important diseases of the central nervous system, he became less supportive. Wisniewski was one of the most flamboyant personalities I had ever met, known to some of his colleagues as a "street fighter," perhaps a reference to his youth in German-occupied Poland. He—and most other scrapie investigators, including Alan Dickinson and Carleton Gajdusek—still believed that a virus caused scrapie and that amyloids were composed of cellular protein, not viral protein, and thus could not be the cause of the disease. Including Diringer and Kimberlin, the five created a formidable opposition.

They mounted a two-pronged argument, contending either that PrP 27–30 and the rods were of cellular origin and thus could not be the virus that caused scrapie, or that PrP 27–30 was a component of the scrapie virus but did not form amyloids, because amyloids were composed of cellular proteins. Their arguments were popular among my antagonists attending the Sixth International Congress of Virology, in Sendai, Japan, in August 1984. By that time, we had produced superb antibodies to PrP, and Steve DeArmond had shown that amyloid plaques in the brains of hamsters with scrapie stained with these antibodies.

Opposition to our finding that PrP 27–30 polymerizes into amyloid should have ended at the Sendai meeting, but it did not. Disturbed by the level of animosity, I invited Alan Dickinson to breakfast at the restaurant in the hotel where the meeting was held. I asked him, since he was the intellectual leader of the field, to try diminishing the hostility, to which he replied, "You must retract the word 'prion' and let me pre-review your papers prior to submission for publication. Then, and only then, shall I become friendlier. Each time you make a discovery, the British authorities cut my budget." When I emphatically replied, "No, the word 'prion' is nonnegotiable and I don't need your approval to publish my results," the breakfast was over. Dickinson got up from the table and walked out.

In the fall of 1984, a paper on SAFs was published in *Science,* jointly authored by Wisniewski, Gajdusek, and their colleagues from Staten Island and Bethesda.[14] They argued that SAFs were distinct from amyloid in the electron microscope and "not just another morphological manifestation of amyloid." Later in the paper, they suggested that "SAF may be the etiologic agent of these diseases. If so, SAF represent a new class of filamentous animal virus." Filamentous viruses, such as tobacco mosaic virus, had previously been seen only in plants and bacteria, never in animals, but the authors reported that besides occurring in scrapie-infected mice, SAFs were also found in crude fractions prepared from the brains of people who died of CJD.

When I read the paper, I remember laughing: I thought it was nonsense. They had ignored the first principle of virus morphology, which

Robley Williams had taught me: "When you've seen one tobacco mosaic virus, you've seen them all" (see figure 20). All filamentous tobacco mosaic viruses look alike in the electron microscope. The virions of a single type of virus are virtually indistinguishable from one another, yet the SAFs pictured in the *Science* paper were of varying sizes. Had the authors looked at Williams's *An Electron Microscopic Atlas of Viruses* or conferred with a microscopist who was an expert on the ultrastructure of viruses?

Diringer, who thought an amyloid-inducing virus caused scrapie, argued that the rods we had seen and Wisniewski's SAFs were identical, and persistently ignored my queries about the morphological differences between SAF and amyloid. As far as he was concerned, the amyloid rods he found in his purified fractions were SAFs, despite the fact that they did not meet the electron-microscopic criterion for classification as SAFs. He would spend the next fifteen years searching for the amyloid-inducing virus of scrapie. He retired from the Robert Koch Institute in the late 1990s still believing that someday, somewhere, someone would find the "scrapie virus" that induces PrP amyloid.[15]

Our discovery of PrP amyloid provoked a firestorm that continued for several years. At an American Society of Neurochemistry meeting in Portland, Oregon, in the spring of 1985, Henryk Wisniewski chaired a session devoted to scrapie at which I presented our work. In the middle of my presentation, Wisniewski interrupted me by holding up his necktie and yelling, "Congo red dye binds to silk."

"Yes, Henryk, you are correct. Silk does bind Congo red dye," I replied.

"But my tie does not cause scrapie," he shouted triumphantly, "and therefore, the scrapie agent is not amyloid."

"Yes and no," I said. "Yes, your tie does not cause scrapie, and no, scrapie prions are amyloid." After this exchange, I resumed my lecture.

Not long after that encounter, I received a letter asking me to give a plenary talk at the upcoming International Congress of Neuropathology in Stockholm. Wisniewski, I later learned from one of the congress organizers, was put out because he had not been invited to give the talk. Despite his protests, the organizers did not back down. Moreover, when I finished my talk, the chair did not permit questions, thus preventing Wisniewski from plastering me with insults. Nevertheless, he rushed from his seat in the audience toward the stage, hollering that SAFs were the cause of scrapie and CJD.

I was not unprepared for Wisniewski's antics. As I recall a conversation with Mike McKinley, he had learned that Wisniewski, Kimberlin, Dickinson, and several others had decided I must be stopped. Apparently, I had received too much credit and the field needed realignment. If the structure of the scrapie agent were ever to be discovered, they wanted the credit spread across the entire field. A plan was hatched: Wisniewski would hire my protein chemist David Bolton and my immunologist Paul Bendheim, and once they were gone, my laboratory would collapse! Their view was that my only talent lay in administration; I had been lucky to hire these stars, and without them I was nobody.

When Bolton and Bendheim each informed me that Wisniewski had offered them jobs, I was not all that unhappy. Both wanted more than I was willing to offer; each wanted to direct a sizable part of the laboratory I had built over the preceding decade. Not surprisingly, I was uninterested in delegating responsibility to either, regardless of how talented they might be. Bendheim posed a particularly difficult problem for me. He wanted to stay in San Francisco, but I could not justify asking Bob Fishman to give him a faculty appointment—from my perspective, he had done little work, beyond helping Ron Barry produce PrP antibodies. However, he had been responsible for my meeting Walter Burke and thus garnering considerable support from the Fairchild Foundation. Had Bendheim's contacts earned him a position even if I found him to be relatively unproductive in the laboratory? I liked Paul—he was warm and charming—but after some thought I decided I couldn't keep him, even if my decision jeopardized the Fairchild support.

As the time approached for Bendheim and Bolton to depart, I organized a large lunch at a superb Chinese restaurant in their honor, at the end of which I made an amiable speech wishing them well. After Bendheim replied in kind, Bolton stood up and said, "This laboratory has much to be proud of, and everything will be fine in the future. Just keep up the good work." To me, the unspoken implications of his toast were that his and Bendheim's departure would sink my laboratory and the epicenter of scrapie research was now moving to Staten Island. I can only imagine the tales of future grandeur the two of them and Wisniewski must have spun.

I can also imagine their frustration for years to come. Despite Wisniewski's pioneering studies, he missed the amyloid/prion connection, and Bolton and Bendheim never produced the breakthroughs their move to Staten Island heralded. At a meeting some years later at Cold Spring Harbor Laboratory on Long Island, Wisniewski confessed to me his displeasure regarding Bolton's and Bendheim's lack of productivity.

10

Finding the Gene

S ecure with the evidence that PrP 27–30 was an essential component of the infectious prion, we needed to learn its genetic origin. By now, I had entered into two collaborations that would
help carry the prion work forward on just that front.

Initially, we considered the possibility that PrP 27–30 was synthesized from a subset of amino acids by a process not utilizing genes. Once
we found that PrP contained all twenty amino acids, the likelihood of a
synthetic process not involving a gene became small. Assuming a gene
encoded PrP, we needed to identify the gene and determine where it resides. Is it part of the animal's genome or is it carried by a virus? Using a PrP
gene probe—that is, a nucleic acid labeled with a radioactive isotope—
we could locate the gene itself and answer these questions. The development of automated protein sequencing and DNA synthesis in Leroy
Hood's laboratory at the California Institute of Technology enabled us to
answer a series of critical questions.

Because prions resisted inactivation by procedures that modify nucleic acids, making the presence of a gene inside them unlikely, I had
wondered whether they were composed of polypeptides of fewer than a
dozen amino acids and similar to certain antibiotics found in bacteria
and fungi.[1] Both bacteria and fungi synthesize nonribosomal peptide
antibiotics without any directions from genes; more than a half-dozen
enzymes work together to assemble these small polypeptides.[2] Some of
the antibiotics bind avidly to fat and membranes, as PrP 27–30 does. I

thus imagined PrP 27–30 as composed of repeating subunits of a dozen or fewer amino acids—but once we found that it contained all twenty of the naturally occurring amino acids, that idea was abandoned.

Developing a gene probe for PrP would not be easy—at that time, the procedures were relatively new. To begin with, we had to determine part of the amino acid sequence of PrP and then use this information to find the gene in a library of cloned DNAs. Fortunately, the path to determining the sequence had been clearly demarcated a few years earlier. In 1981, I met Lee Hood at a Howard Hughes Medical Institute conference in Miami, where he described the automated sequencing of proteins. Using chemical procedures to chop off amino acids one at a time from the end of a protein, Lee had built an instrument that automated what was previously a laborious process at best.

Before I describe the workings of Lee's instrument, it is important to paint a picture of how proteins are constructed. They are chains of amino acids that are assembled inside cells beginning at what is called the N-terminus. By convention, scientists place the N-terminus on the left and the C-terminus on the right. Lee's instrument contained a reaction chamber, where amino acids were cleaved from the N-terminus one by one and identified. This operation was cyclical, repeated over and over until as many as thirty consecutive amino acids (out of, say, one hundred) had been determined. Not only was this instrument much more reliable than any manual process, but its increased efficiency also permitted longer stretches of a particular protein to be sequenced. However, the purity of protein to be sequenced remained crucial: If the protein you hoped to sequence was 60 percent of the total sample and two other proteins represented 15 and 25 percent each, you would be in trouble. Assuming that each of the three had a different amino acid at the N-terminus, you would probably see three different amino acid signals during the first cycle. Multiple amino acid signals during each cycle of the sequencing process would prevent you from determining the sequence of your protein.

A forceful and optimistic scientist, Lee made the process sound simple (figure 22). After my presentation on the purification of the scrapie agent, he said he would be glad to determine the amino acid sequence of the scrapie protein, once we had isolated it. Therefore, even before we published the two papers describing the discovery of PrP 27–30 in December 1982, I had asked David Bolton to focus his energy on preparing

Figure 22 Darlene Groth (see figure 11) prepared purified prion samples for sequencing by Steve Kent, who was working in the laboratory of Lee Hood (above).

the protein for sequencing. To achieve the necessary level of purity, we had to develop one or two clever steps for removing the few remaining contaminants. I could see that Bolton was not thrilled about doing this kind of cleanup work, so I arranged a trip to Caltech, where Lee Hood showed him the sequencing instrument. This firsthand look at a stunning new technology would be enough, I hoped, to spur Bolton into action, but over the next few months, and during my prolonged bed rest in the winter and spring of 1983, he spent little time on the purification project. Frustrated by his reluctance, I told him that the project might progress better if I assigned Darlene Groth to these studies. Bolton raised no objections, and Darlene took on the work with her usual exuberance for critical assignments.

The aggregation of PrP 27–30 into amyloid rods was at the heart of the purification problem. Contaminating proteins were so tightly bound to the rods that we needed to disrupt the rod structure to remove them. But the harsh chemicals needed to break up the rods also denatured (unfolded) the PrP 27–30, resulting in a loss of prion infectivity. We were between a rock and hard place: Each time we disrupted the rods, we lost the only way of measuring whether we had enhanced our purification. I decided we had to take some risk in order to move ahead. Since by now we thought that the 27-to-30 kDa PrP 27–30 represented at least 80 percent of the protein in our preparations, we could probably afford to denature the protein and then enhance purification simply by using gel electrophoresis to separate the proteins according to size alone.

Beginning in the fall of 1983 and continuing into the next spring, Darlene made five pilgrimages to Lee Hood's laboratory in Pasadena with our most purified preparations. Darlene worked closely with Hood's colleague Stephen Kent, who was in charge of the sequencing efforts, but at first the results were disappointing. Each time a new preparation was sequenced, the data were uninterpretable, with two or three different amino acids found in each sequencing cycle. Could our protein preparations still contain more than one specific protein? The puzzle was solved by Steve Kent, who discovered that the explanation for the apparent multiple sequences was a ragged N-terminus: The limited protease treatment we found so effective in purifying scrapie infectivity had resulted in the digestion of PrP at slightly different places on the terminus.

We now had only one amino acid sequence: that of PrP 27–30. Based on the sequence, Lee Hood produced DNA probes that could retrieve the PrP gene. He sent them to me, but although I knew what needed to be done, I didn't have the expertise in my laboratory to take the next step, which was to identify clones of the PrP gene.

Interlude: Deciphering the Amino Acid Sequence of PrP 27–30 and Pursuing the Gene

Each time Darlene brought new purified preparations of PrP 27–30 to Pasadena, the sequencing data were uninterpretable, with two or three different amino acids found in each sequencing cycle. Apparently, our protein preparations still contained more than one protein. Darlene and I were both becoming discouraged, but we had no choice except to persevere. With Kent and Hood, we decided on two new approaches: First, we sequenced PrP 27–30 that had been denatured and purified by a slightly different procedure; and second, we injected infectious fractions containing the amyloid rods directly into the sequencing chamber. Both experiments gave the result we had seen earlier—that is, in each cycle of the sequencing process, the same two or three amino acids cropped up. We were genuinely puzzled and began to wonder whether our interpretation of the data was correct. Perhaps our preparations were sufficiently pure and there was another explanation for the apparent heterogeneity.

There were thousands of possible sequences that could be pieced together from various amino acids found in each degradation cycle. We needed a way to simplify the analysis, an approach that would reduce the complexity of the data. It came when Steve Kent divided the amino acids found in each of the degradation cycles into three categories based on their relative abundances: The amino acids in the first group were in high concentration, those in the second group were intermediate, and those in the third group were low. This categorization of amino acids by abundance turned out to be a giant leap forward. Next, Steve wrote out an amino acid sequence defined by the most abundant amino acids in each cycle. Then he repeated this process to come up with a second sequence, comprising those amino acids with an intermediate abundance, followed by a sequence made up of those with the weakest signals. Immediately, Steve saw the solution.

When Steve moved the sequence of intermediate abundance two positions to the right, it lined up perfectly with the sequence made up of the most abundant amino acids. Next, he moved the least abundant sequence four positions to the left and again found perfect alignment. We now had only one amino acid sequence: that of PrP 27–30! The explanation for the apparent multiple sequences was a ragged N-terminus: The limited protease treatment we found so effective in purifying scrapie infectivity had resulted in the digestion of PrP at slightly different places on the N-terminus.

In possession of a unique amino acid sequence for PrP 27–30, we proceeded to retrieve complementary DNA (cDNA) clones encoding PrP. To do this we employed recombinant DNA techniques that had begun to revolutionize the biological sciences. To produce a gene library, the mRNA is reverse-copied into DNA again, using an enzyme called reverse transcriptase. Since cDNAs are reverse copies of messenger RNAs, they are relatively small and do not contain any junk DNA sequences. Once created, the cDNAs are introduced into bacteria, where they are amplified—that is, produced in great numbers. Each bacterium, as it replicates and grows into a large colony, will make many copies of only one cDNA. The DNA probes that Lee Hood made would allow us to identify which of these colonies produced the cloned cDNA that encoded PrP 27–30.

Although we had a path to the gene through a unique amino acid sequence, I was worried that others might pass us. I intended to present the amino acid sequence at upcoming meetings and publish it—thus others would soon have the same key to unlocking the mysteries of the PrP gene. Despite gene cloning having been developed in part at UCSF, no one with the appropriate experience there expressed an interest in helping us; it may be that my UCSF colleagues were still leery of the idea that proteins could mimic viruses by reproducing themselves and causing infection. Unclear about how to proceed, I considered turning the cloning process over to Lee Hood, but a few friends cautioned me that the cloning of the PrP gene might get lost in his laboratory, which now numbered more than a hundred scientists working on such things as the immune system, better sequencers for protein and DNA, and improved DNA synthesizers.

In the summer of 1982, at an International Biochemistry Congress in Perth, Australia, I had met the Swiss molecular biologist Charles Weissmann, who was renowned for having cloned the genes that encode interferons, a family of proteins effective in treating viral infections and possibly cancer. After my talk, Weissmann, then based at the Institute for Molecular Biology in Zürich, proposed coming to my laboratory on a sabbatical; he hoped to apply some of the clever techniques used in his interferon work to find the putative nucleic acid genome of the scrapie agent that, he felt, I must have overlooked. I was thrilled at the possibility that Weissmann might come to San Francisco and spend a sabbatical—there was so much I could learn from him. He had an engaging smile that conveyed his warm personality, yet I found that he could be severe and judgmental as well. He wore thick-rimmed glasses, which he removed when reading small print. His Talmudic approach to science made him a wonderful critic; he was very smart and extremely analytical (figure 23).

The sabbatical never transpired, but the following year Weissmann came to San Francisco to present his interferon studies at the annual meeting of the American Society for Biochemistry. We had lunch and talked about the relative merits of the RNA and protein approaches to scrapie. I gave him some of our most highly purified prion preparations, and he also set up a series of experiments with one of his graduate students to compare RNAs prepared from scrapie-infected and control brains of hamsters. My attitude toward his search for an essential RNA was similar to what I had expressed to my postdocs who thought they had evidence for a scrapie-specific nucleic acid: More important than

Figure 23 Charles Weissmann (above) and Bruno Oesch determined the DNA sequence of the PrP gene in Zürich.

proving the absence of a nucleic acid was uncovering the truth about the scrapie agent.

Since by now a year had passed and Weissmann had not yet identified an RNA that was unique to scrapie, I thought he might be willing to screen his hamster gene libraries in a search for the PrP 27–30 gene using the DNA probes Lee Hood had synthesized. At the same time, I was afraid he might try to take over the problem. Some of my friends suggested that Weissmann might trample me if he decided the problem was worthwhile. But after a few weeks of deliberation, I settled on Weissmann as the man to do the cloning and sent the probes to Zürich.

Pariah and Prophet

In July 1984, with a paper in which Lee Hood and I described the sequencing results accepted for publication in *Cell*,[3] I boarded an airplane bound for Edinburgh. I was on my way to a conference whose organizers, I knew, had attempted to exclude me while including my data. Almost a year earlier, at a meeting at the Rocky Mountain Laboratories, Richard Kimberlin had approached me and, looking rather sheepish, said, "Alan Dickinson wants your data presented at a meeting next summer in Edinburgh, but he'd rather not have *you* there." I was taken aback, to say the least. How could Kimberlin have mustered up the audacity to say this to me? "Richard," I replied, "the only way you and Dickinson will get my data is to invite me to the meeting"—and so the invitation was sent, apparently at the instigation of Kimberlin and Heino Diringer.[4]

When I arrived, I could feel hostility oozing from the very stones of Newbattle Abbey College, where the meeting took place. The abbey dates from the sixteenth century, and the accommodations were correspondingly ascetic, down to a lack of telephones in the rooms. Seeing Erling Norrby again was the one bright spot. Erling and I had become friends when he spent a sabbatical at UCSF in 1977–78. He is one of the most politically correct people that I know; always appropriate, gentle, and patient. Quite knowledgeable about scrapie and kuru, he had been a willing and informed critic. We had spent many hours discussing the need to purify the scrapie agent and its physical behavior in the centrifuge.[5] By the time of the Edinburgh meeting, of course, I was no longer forced to present studies on the scrapie agent's sedimentation properties. My talks this time were on the prion amyloid and the N-terminal sequence of PrP 27–30.

Of the nearly one hundred people at the meeting, Erling was the only one friendly to me. I was the embodiment of the word "pariah"; the antagonism was palpable. After my talk on prion amyloid, Kimberlin polled the audience as to whether the amyloid rods composed of PrP 27–30 were SAFs, and everyone he called on agreed that the rods and SAFs were the same (some of them among those who would argue a month later, in Sendai, that SAFs were filamentous viruses and unrelated to any amyloid). No one seemed able to conceive of a scrapie virus that contained an amyloid protein much less an amyloid prion that caused scrapie. In their minds, there was an inviolable separation between viral proteins and the cellular proteins that polymerize into amyloid. When Kimberlin finished his polling, he turned to me and asked (as if the sheer number of people disagreeing with me would serve to change my mind), "Now, don't you agree that the rods and SAFs are one and the same?"

"No," I said. "The rods have been shown to be amyloid by electron microscopy and by Congo red dye staining. The rods are composed of PrP 27–30. The SAFs have been repeatedly shown to differ from amyloid—morphologically and tinctorially. Furthermore, SAFs have not been purified, so we don't know the composition." This sally, however, was met with an eerie silence, and the session broke up.

My presentation of the N-terminal sequence of PrP 27–30 the following day should have been the highlight of the meeting. The determination of the sequence meant that all the wonderful new techniques of molecular biology could be applied to unraveling the mysteries that still shrouded prions, but few of the investigators wanted to acknowledge our achievement. After my presentation, Konrad Beyreuther, a biochemist who was collaborating with Heino Diringer, came up to me and announced, "Heino and I just finished sequencing PrP 27–30, and your sequence is wrong." His words struck me like a bolt of lightning. Our paper on the sequence of PrP 27–30 would soon be published in *Cell*, unless we told Ben Lewin that we were unsure of the data. I quickly found a pay phone and called Lee Hood in Pasadena: "Lee, I'm at a scrapie meeting in Edinburgh, and I was told a few minutes ago that our amino acid sequence of PrP 27–30 is wrong."

"Our sequence is fine," Lee responded. "But I can look at it again with Steve Kent. We have lots of data."

Still agitated, I said, "Lee, please go back over the data. I'll phone you tomorrow. You can't reach me; there are no telephones in the rooms." After dinner with Norrby in a quaint Scottish seafood restaurant near Edinburgh Castle, I went back to my room and spent a restless night in the grip of anxiety. When I phoned Lee the next day, he told me that he and Steve had carefully reviewed the data and were certain the sequence was correct. I was greatly relieved and eager to tell Beyreuther that it was he and Diringer who were mistaken, not I. When I did so, after one of the sessions an hour or so later, he seemed unperturbed, a reaction that made me wonder whether he had any data at all—was he playing with my head? Had he merely tried to keep us from publishing our data before he and Diringer had a chance to publish theirs?

By the end of the weeklong meeting, I was exhausted; I just wanted to leave. I was due to confer with Charles Weissmann before returning to the United States, and when the plane took off from Edinburgh en route to Zürich, I felt as though I had been liberated from a Vietcong POW camp. My brainwashers were deeply troubled by the data we were obtaining (the data they had wanted at the meeting without me there to present them). I was supposed to be finding a virus, and failing to find one was inexcusable. There had to be something wrong with me!

Weissmann met me at the Zürich airport in a great state of excitement. "PrP is encoded by a chromosomal gene, not a viral gene!" he announced. He and his student, Bruno Oesch, had used the DNA probes I sent to identify a gene in a cDNA library that had been prepared from messenger ribonucleic acid isolated from the brain of a normal, uninfected hamster. I was thrilled with his success—it had all happened so fast.

We drove from the airport to the rather drab Hönggerberg campus of the fabled Eidgenössische Technische Hochschule (ETH), Einstein's alma mater and home of the Institute for Molecular Biology, which Weissmann had founded. The campus, spread over more than a dozen acres, consisted of a clutch of concrete rectangular buildings some four or five stories high. From an underground parking garage, we climbed a grim flight of stairs and entered a long corridor lined with laboratories. Halfway down was Weissmann's modest, workmanlike office, where Bruno Oesch joined us.

Hood's probes had identified the cDNA copy of the PrP gene among millions of unrelated cDNAs. When Weissmann found the N-terminal amino acid sequence of PrP 27–30, he knew they had isolated the correct gene. Curiously, the initiation signal for producing a protein seemed to be missing: Generally, the first amino acid of a protein is methionine, which is encoded by the DNA base sequence ATG. But Weissmann could not find an ATG at the beginning of his DNA sequence, forcing us to conclude that PrP 27–30 is cut from a larger protein. This was consistent with our explanation of the ragged N-terminus of PrP 27–30, which had been so problematic until Steve Kent deciphered the riddle. More to the point, Weissmann had recovered a DNA clone that encoded PrP, *but the clone had been produced from an uninfected hamster.* (Several years later, I learned that the hamster had been hurriedly bought at a pet store in

Zürich and not from a supplier of laboratory animals as I had believed, but luckily it didn't matter.) I had to conclude that the PrP gene was present in normal animals; there seemed to be no other interpretation.

How could the gene for a protein causing scrapie reside in healthy animals (and presumably people)? "Do I have a PrP gene?" I began to wonder.

I needed to be absolutely sure of the genetic origin of PrP. Was PrP encoded by a cellular gene, as Weissmann claimed, or was it the product of a foreign gene, such as one carried by a virus? Using the cloned DNA that Weissmann had retrieved, David Westaway, who had recently joined my group as a postdoc after several years at UCSF working with Harold Varmus, found that DNA isolated from both uninfected and scrapie-infected hamster brains had a PrP gene. These findings forced us to conclude, as Weissmann had, that the PrP gene is contained in the animal's own genome and not carried by the mythical scrapie virus. Next, David looked for a PrP gene that might be buried inside the infectious prions. Using highly purified preparations, he searched for a PrP gene but was unable to find even a trace.

PrP gene-expression studies in Zürich and San Francisco gave the same results: The levels of messenger ribonucleic acid encoding PrP were similar in normal and in scrapie-infected animals. How could this be? Was PrP harmless? Had we isolated the wrong protein? Maybe scrapie infectivity depended on another protein. It was easy for me to imagine Hood and Weissmann shaking their heads and mumbling that Stan, poor guy, had spent the last eight years purifying the wrong protein.

Had I? I set out to answer that question by asking Lee Hood to perform additional sequencing studies on pieces of ultrapure PrP 27–30. When we found that the sequence of each fragment of PrP 27–30 was represented in the gene sequence, I became certain that at least we had retrieved the right gene. Still, despite many hours of pondering the meaning of PrP messenger ribonucleic acid expression in uninfected animals, I was stymied, back in the thick Pacific coastal fog. I needed a path through that fog. I had to find some explanation other than that we had isolated the wrong protein. If PrP 27–30 was unrelated to the infectious scrapie prion, my scientific career was toast. Almost a decade of work—long hours in the lab, tomes of proposals seeking funding, thousands of animals sacrificed—would be reduced to nothing.

I decided to focus on finding the protein produced from the PrP messenger ribonucleic acid in a normal, uninfected brain. I suggested to Ron Barry that the most unusual feature of PrP 27–30 was its resistance to degradation by proteases and that PrP in normal cells might not have this unusual property. Ron homogenized brains from uninfected hamsters in the presence of protease inhibitors—molecules that prevent the degradation of proteins. He then used the antibodies he had raised against PrP 27–30 to look for a similar protein in these normal brain homogenates. When the antibodies stained a protein that was slightly larger than PrP 27–30, the explanation began to emerge.

Ron had discovered a new protein, which we called PrP^C, "cellular PrP." The superscript "C" distinguished it from a protein of the same size found in scrapie-infected brains, which we designated PrP^{Sc}, for "scrapie PrP." Upon exposure to proteases, PrP^C was destroyed altogether, whereas PrP^{Sc} was merely truncated, to form PrP 27–30 (see figure 21). It seemed likely that the N-terminal region of PrP^{Sc} was cleaved during this protease treatment, which explained the smaller size of PrP 27–30, its ragged N-terminus, and the missing methionine at its beginning.

Now that pieces of the puzzle were coming together, I started to feel better. These findings changed our thinking dramatically and allowed the work to accelerate. The information on the prion protein had surfaced in a rather disordered fashion, but this is characteristic of most major discoveries, especially those that force changes in how we think. Often there is initial chaos, but after much thought, a realignment of the findings presents a new synthesis, and a novel understanding emerges. The fog cleared with the discovery of PrP^C and PrP^{Sc}, and a path forward opened. I could now see far ahead across the landscape of our discovery.

Our findings presented a fascinating new set of questions, but before launching into new studies, we had to publish our work on the PrP gene, its expression, and the discovery of PrP^C and PrP^{Sc}. I wanted to assemble all the data into one paper, but I had second thoughts when Weissmann insisted that his graduate student, Bruno Oesch, be the first author and that he be the senior author, whose name would appear in the last position. Eventually, I decided that one superb paper describing these advances was more important than the order of the authors.

Assembling the manuscript was not easy, even though all of the figures, except for the sequencing, had been prepared in San Francisco.

Weissmann, a man used to getting his own way, wanted every sentence filled with caveats; he still believed that the mythical nucleic acid of the prion would someday be found, and he missed no opportunity to insert this view. Negotiating the wording of a paper with him that extended the prion story and at the same time exuded caution was no small task. After many hundreds of dollars spent on telephone calls between San Francisco and Zürich, we finally had a manuscript, which we sent to Ben Lewin for publication in *Cell*.[6]

Soon after our paper was published in the April 1985 issue, a paper appeared in *Nature* by Bruce Chesebro and his colleagues at the Rocky Mountain Laboratories, concluding that PrP 27–30 could not be an essential component of the infectious scrapie agent, since the gene was expressed at equal levels in both scrapie-infected brains and uninfected controls.[7] Because Hood and I had published the N-terminal amino acid sequence in *Cell* the year before, anyone with experience in gene cloning could retrieve the PrP gene. Chesebro had read our 1984 paper and decided to use the sequence to synthesize his own set of DNA probes. He then used the probes to identify the PrP gene from a library he had produced from mouse brain. Much later, I learned about the celebration this finding prompted at the Rocky Mountain Laboratories, particularly when it was presented to the Board of Scientific Counselors during a review of Chesebro's work. One of the counselors described to me the elation and apparent satisfaction that Chesebro and others exuded as they presented their data arguing that I must be wrong. Chesebro, whose work on persistent viruses, including HIV, was superb, would remain trapped in the world of virology. Having invested so much time and effort looking for a virus, he would continue to believe that a virus caused scrapie many years after most scientists had accepted prions as the real cause.

It seems I hadn't grasped just how nervous Weissmann was about our *Cell* paper. Once it was published, Ted Diener asked me, "What in the world are you doing with all those caveats?" Nevertheless, my collaborations with Hood and Weissmann filled him with admiration. "Most people would be so scared of being run over," he said, "that they wouldn't dare collaborate with such renowned scientists." In retrospect, I can see that I was naïve on that score—but I had needed help. My naïveté saved me: I could not possibly have mastered all the disciplines required for

Figure 24   Detlev Riesner and I expended an enormous effort to find evidence for a virus or a small nucleic acid hiding inside the prion.

determining the nature of the prion. The more outstanding the scientists I could entice to work on scrapie, the more likely it was that a solution to the problem would emerge.

Hood and Weissmann, for their part, reacted to this bonanza of new biology in quite dissimilar ways. Hood was happy to continue helping us open up this new frontier, but he was too busy to become intimately involved. His work on the immune system was moving forward rapidly, and so was his development of the extraordinary instruments for protein and DNA sequencing. Weissmann remained convinced that prions must have a small nucleic acid hiding somewhere, and he and I spent many hours discussing how prions could possibly replicate without one. His Talmudic approach to science and his penetrating formulations of problems were enlightening—if sometimes a little ponderous. He was relentless but persuasive in suggesting experiments I might do in collaboration with Detlev Riesner, an expert on the physical chemistry of viroids at Heinrich Heine University in Düsseldorf, to find the mythical nucleic acid within the prion. Detlev has a gentle, engaging face that portrays his kind and affable personality (figure 24). When we first met, his straight

Figure 25   With George Carlson (top left), we mated ILnJ mice that had long prion incubation times with short-incubation-time NZW mice. All the (NZW × ILnJ) F1 offspring of this cross had long incubation times similar to those of the ILnJ parent. To determine whether the long incubation times in F1 hybrid mice were due to the effects of a single prion incubation time gene, we mated the F1 hybrids to short-incubation-time NZW mice. Approximately half of the offspring of this backcross had short incubation times and the other half had long ones, compatible with the major effect of a single gene. Strikingly, fifteen mice that inherited the NZW allele of the PrP gene (a/a) from their F1 parent had short incubation times (upper graph), while fifty-one mice that inherited the ILnJ allele (a/b) had long incubation times (lower graph). Later, David Westaway (top right) found that the PrP genes of NZW and ILnJ mice encoded proteins that differed by two amino acids.

brown hair parted on the left framed his wide forehead that rose above his warm blue eyes. Despite not seeing each other very often, Detlev and I became close friends. My collaboration with Detlev in a vain search for an essential RNA buried inside the prion was to last nearly twenty years.

Genetic Control of Incubation Times in Mice

Discovery of the PrP gene had opened up a fertile line of questioning. For example, we could ask, "Does PrP control the onset of scrapie?" A couple of years before clones of the PrP gene became available, I had begun a collaboration with George Carlson, a geneticist at the Jackson Laboratory in Bar Harbor, Maine, a leading mouse genetics research laboratory. George was a tall man with a thin face crowned by straight brown hair parted on the left side. Though shy and quiet, this self-effacing man was a brilliant scientist whose modesty almost always was inappropriate (figure 25). We wanted to know whether the PrP gene had a role in regulating the length of the scrapie incubation time. With a probe for the PrP gene, we could now examine strains of mice with long and short incubation times.

In the late 1970s, I had asked one of my postdocs to order about a dozen immunologically compromised mice from the Jackson Laboratory so that we might determine incubation times for the scrapie agent. Might there be variations in lines of inbred, commercially available mice that would offer a new avenue of investigation? We found one mouse line— New Zealand White (NZW)—with incubation times of about 110 days and another line, ILnJ, with incubation times exceeding 200 days.[8] No longer did I need Alan Dickinson's VM/Dk mice, with their long incubation times, which he continued to hoard. We had our own lines with various incubation times, and we could begin to unravel the incubation-time conundrum. Using the VM/Dk mice, Dickinson and his Edinburgh colleagues had reported in 1968 that a single gene controlled the length of the incubation time and that long incubation times were dominant traits.[9]

With our own mouse lines, which had different incubation times, we could begin to unravel the incubation-time conundrum. We started by crossing the NZW with the ILnJ mice. Similar to Dickinson's results published in 1968, all of the offspring denoted (NZW × ILnJ) F1 had long incubation times. But these findings made no sense to me. The F1 off-spring from the mating of NZW and ILnJ mice carried one gene copy for short incubation times and another for long ones. Why was the protein encoded by the short-incubation-time gene unable to support rapid prion replication and produce a fatal disease after an abbreviated time? It would take another six years for George and me to decipher the mechanism by which long incubation times are dominant.[10]

Interlude: Strategy for Breeding Experiments Using Genetic Crosses

In the mid-1980s, George Carlson and I set out to determine whether the incubation time difference between NZW and ILnJ was due to a single gene. The first genetic cross of the mice was denoted (NZW×ILnJ) F1. The term F1 refers to the first filial generation after the parents. The incubation time results in the F1 mice dictated how we set up the next cross to determine whether incubation time was a monogenic trait. There were three possibilities: first, the F1 hybrid mice had short incubation times similar to NZW, in which case the next step would have been to backcross the F1 to ILnJ mice. The second possibility was that the F1 hybrids had long incubation times similar to ILnJ, in which case the next step would have been to backcross the F1 to NZW mice. Third, if incubation time were intermediate between the two parents, we would have done an F2 intercross by breeding pairs of F1 mice. In such experiments, it was critical to use inbred strains of mice in which both copies of each gene are identical.

Since incubation times in the F1 hybrid mice were similar to those of the long-incubation-time ILnJ parent, we chose to cross F1 hybrid mice back to the NZW short-incubation-time parental line. This backcross had the greatest power to determine whether a single gene could explain the incubation time phenotype. The mean incubation time for twenty NZW mice was 113 days while twenty-one ILnJ mice had an incubation time of 255 days; twenty-four (NZW × ILnJ) F1 mice succumbed after 223 days. For the [(NZW × ILnJ) F1 × NZW] backcross, we bred one hundred and thirty mice: sixty-four exhibited a mean incubation period of 130 days and sixty-six an incubation time of 195 days. This was excellent evidence for a single gene controlling the scrapie incubation time, just as Dickinson had found.

When the incubation time experiments using the mice from the back-cross were already under way, the PrP gene probe became available, and we asked if there was a difference between the PrP genes of the NZW and ILnJ mouse strains. Once we found a difference that allowed us to track the PrP genes from NZW and ILnJ parents through the two sequential crosses, we collected tail snips from all the living mice for DNA analysis.[11]

Using the PrP probe, we found that all the mice in the F1 cross had one copy of the PrP gene from the NZW parent (designated the "a" allele) and one copy of the PrP gene from the ILnJ parent (designated the "b" allele), as expected. Next, we examined the DNA isolated from the tails of the mice in the backcross: fifteen mice carried only the PrP "a" allele that was inherited from the NZW mice and all of these mice exhibited short incubation times (see upper graph in figure 25). Fifty-one mice carried both the "a" and "b" alleles and all but one discordant mouse showed long incubation times (see lower graph in figure 25).[12] The relatively small number of mice in the group carrying the PrP "a" allele exclusively was due to the deaths of many of these mice prior to the time that we began collecting tail snips.

Our excitement was palpable. Starting with crude extracts of brains from scrapie-infected mice, we had isolated a single protein that transmitted disease when inoculated into naïve mice. Then through an arduous series of experiments, we managed to sequence a segment of the PrP protein and retrieve the corresponding gene. Next, we had been able to use a PrP gene probe to show that either the PrP gene itself or another nearby gene controlled the length of the scrapie incubation time in mice.

Somewhat later, David Westaway made a stunning discovery: The amino acid sequence of PrP in the ILnJ mice differs at two positions from PrP found in NZW mice with short incubation times,[13] as determined earlier by Bruce Chesebro. This finding strongly supported the idea that PrP itself controls the length of the incubation time.

Even though we still did not know the mechanism whereby the prion protein reproduced and became infectious, it seemed increasingly unlikely that the prion story would fall apart. I felt more and more certain that we had not misinterpreted our data. Thirteen months later, the Edinburgh group would confirm our genetic-linkage studies, using the VM/Dk mice that Dickinson had steadfastly refused to share.[14] The stress was beginning to dissipate. The sky looked as though it might not fall down anytime soon.

11

Jousting with the Press

A lthough I had already had more than my share of publicity, I considered it my duty to speak to interested reporters. How else would the public learn about our work, and more important, about scientific research in general? It is a widely held belief that the more informed the public is through newspaper and popular magazine articles, the more supportive they will be of scientific research. But sometimes these attempts backfire.[1]

A couple of years before David Perlman's original story about prions appeared in the *San Francisco Chronicle,* a reporter from *Omni* magazine came to see me. He was enthusiastic about the systematic centrifugation studies we had undertaken with the aim of solving the mystery of what causes scrapie. When the *Omni* article appeared, I was described in glowing terms as a "carpet salesman."[2] I was mortified! I immediately called the reporter, who said that "carpet salesman" was not his metaphor—the editor had inserted it at the last moment. When I called the editor, she explained that the day before editing the article she had visited a rug bazaar, where the salesman was so nice and kind that she thought she could capture the reporter's enthusiasm for me and my work by referring to me as a carpet salesman. When I pointed out the derogatory parallels—to me, the image of most carpet salesmen is no better than that of most used car salesmen—she admitted I had a point and apologized. I had spent several hours explaining our work to the *Omni* reporter, and the result was an insulting comparison to a slick salesman from a souk.

In December 1986, shortly after George Carlson and I had published our genetic-linkage study in *Cell,* a rather more defaming article appeared in *Discover* magazine. The article, by the science writer Gary Taubes, was entitled "The Name of the Game Is Fame, but Is It Science?"[3] and its tenor may be gauged by its title. Taubes had smelled a good story. My work was progressing well, and support from the NIH and the Fairchild Foundation was propelling it forward. My competitors were becoming increasingly resentful. The prion train had left the station, and the scrapie-virus hunters were still standing on platform, shouting "Virus!," "Virino!," and "Nucleic acid!" as loud as they could. They had no data to support their predilections, but their cries did not go unheeded.

Taubes and his editors at *Discover* preferred to focus on the controversy rather than on an evolving set of scientific findings. Apparently, digesting a series of complex experimental data and distilling it into a captivating article was more work than they were prepared to do. The article opened with an epigraph, a quotation from me: "'Prion' is a terrific word. It's snappy. It's easy to pronounce. . . . It isn't easy to come up with a good word in biology." A searing indictment of me and my science followed:

> Naming something before you discover it is a risky business. For starters, you have to couch your definition with great care, so that it will fit whatever it is that you find, when and if you actually find it. . . . You have to hope that as the data come in they fit your definition, or at least that your definition can be made to fit the data.

Quickly, Taubes segued into the press's love affair with the term "prion" and noted my "flair for public relations." He ignored the caveats I had laid out when I introduced the word four years earlier. His recitation of the science in the published papers by my competitors and me over the previous four years suggested to me that he had not read them closely, if he had read them at all. In that interval, I had reported the discovery of PrP 27–30, its N-terminal sequence, and its ability to polymerize into amyloid. The staining of amyloid plaques with antibodies to PrP 27–30 was considered a breakthrough in medical science. I expected the community of scrapie researchers to be thrilled with the gene-cloning studies showing that a cellular gene encodes both PrP 27–30 and the precursor protein PrP^C. And I would have thought that any science writer would find our genetic-linkage studies, showing that the PrP controls the scrapie incubation time in mice, to be most persuasive. I was wrong.

Taubes depicted the naming of the prion as little more than a public relations ploy. But two-thirds of the way through the 1982 *Science* paper, in a section entitled "Novel Properties of the Scrapie Agent," I had written,

> The foregoing summary of experimental data indicates that the molecular properties of the scrapie agent differ from those of viruses, viroids, and plasmids. . . . Because the dominant characteristics of the scrapie agent resemble those of a protein, an acronym is introduced to emphasize this feature.[4]

Before offering the acronym and defining it, I argued that it was time to retire the word "virus" in this context, with and without its various modifiers. To me, there was so much evidence against a virus causing scrapie that it seemed misguided to continue using the word. If people didn't like my acronym, they could use the word "agent." In the same paragraph, I wrote: "In place of such terms as 'unconventional virus' or 'unusual slow virus-like agent,' the term 'prion' . . . is suggested. Prions are small *protein*aceous *in*fectious particles which are resistant to most procedures that modify nucleic acids."

I continued with several caveats:

> The term "prion" underscores the requirement of a protein for infection; current knowledge does not allow exclusion of a small nucleic acid within the interior of the particle.
>
> Our data and that of others suggest two possible models for the scrapie agent: (i) a small nucleic acid surrounded by a tightly packed protein coat or (ii) a protein devoid of nucleic acid, that is, an infectious protein. While the first model might seem the most plausible, there is no evidence for a nucleic acid within the agent. The second model is consistent with the experimental data but is clearly heretical. Skepticism of the second model is certainly justified. Only purification of the scrapie agent to homogeneity and determination of its chemical structure will allow a rigorous conclusion as to which of these two models is correct.

There seems to be little advantage in championing one model over another; however, several previously postulated structures for the scrapie agent can now be discarded.

Taubes, however, declared that "[Prusiner] hadn't simply put forth what he might then refer to as his 'prion model.' . . . Prusiner had unequivocally and pre-emptively renamed the very subject of his research, as well as that of a few dozen other scientists. From then on, he was no longer examining the nature of slow viruses; his work was about prions."

On the contrary, in the three paragraphs from my paper quoted above, I used the word "model" six times. In the paper's concluding section, I reemphasized my caution:

> The consequences of understanding the structure, function, and replication of prions are significant. If prions do not contain a nucleic acid genome which codes for its protein (or proteins), alternative mechanisms of replication and information transfer must then be entertained. . . .
>
> The importance of prion research in the elucidation of a wide variety of medical illnesses underscores the need for purification of the scrapie agent to homogeneity and the subsequent identification of its macromolecular components. Only then can we determine with certainty whether or not prions are devoid of nucleic acids. Indeed, recent progress in scrapie research has transformed an intriguing yet forbidding problem into an exciting and productive area of investigation.

In other words, I had carefully stated that I could not distinguish between a protein as the sole component of the prion and a protein with a small nucleic acid. Moreover, in the ensuing years, whenever I was asked about the possibility of the scrapie agent turning out to be a virus, I repeatedly said that if a large nucleic acid were to be found and the scrapie agent was shown to be a virus, then the word "prion" should be discarded.

Many of my colleagues might have testified to my circumspection; however, in his choice of people to interview, Taubes managed to corral some of the most disaffected. He quoted Paul Bendheim ("Prusiner 'rammed that word down the throats of everybody in that laboratory and in the world'"); David Bolton ("Even if the scrapie agent was a virus and all the data was staring him in the face, my bet would be that he'd never see it"); and Richard Kimberlin ("The biology of scrapie tells us the damn thing has a genome. . . . To try to create models of an infectious protein with which you can encode strain variation is damn difficult. And Stan Prusiner, bless his heart, never even tried—which is wise, because you can't do it"). The sole quotation that created a little balance in Taubes's otherwise lopsided survey came from an unnamed scientist at the Albert Einstein College of Medicine in New York, who noted of a critique of my work in a letter to the *New England Journal of Medicine:* "Her letter was vicious. It's typical of these people. They hate him. They're watching him run away with what they spent their life trying to do—and failing to do."

Needless to say, I felt betrayed by Taubes, with whom I had spent a lot of time discussing my research. I clearly had not convinced him that our studies were likely to open a new area in biology and medicine and that they were unlikely to be overturned. His piece portrayed me as a marginal scientist whose work would eventually be shown to be wrong. Ironically, the article was so derogatory that many scientists outside the scrapie field became much stronger supporters of my work.

Julie Krevans, Walter Burke, and Fred Seitz were supportive during that trying period. Julie told me to ignore the garbage. "Let me tell you a story about an encounter I once had with the press," he said. "When I was head of the Johns Hopkins medical services at Baltimore City Hospital, an article appeared in the local newspaper about the high death rates among older people who were hospitalized in our facility. I was upset, and so I looked up the death rates among hospitalized older people across the United States and found that our rates were similar. I wanted to share this information with the reporter, so I asked him to meet with me. When he came to my office, I gave him copies of all the epidemiological articles I had consulted. The next day, the newspaper carried a major article entitled 'Head of Baltimore City Hospital Admits to High Death Rate Among Elderly Patients.'"

Nervous about how the *Discover* article would be received by the board of directors of the Fairchild Foundation and aware that they would soon hear about it, I sent a copy to Walter Burke. "At a time when you're down," he responded, "we need to continue supporting your efforts. If there is something to this article, then the board will reconsider our support at a later date."

Fred Seitz was upset, as was Charles Weissmann; both sent letters to *Discover* protesting the assassination of my character. Only Weissmann's letter was published,[5] perhaps because he raised the possibility that my introduction of the word "prion" was premature:

> I read your article with considerable dismay, because its thrust is not to deal with an extremely interesting scientific subject, namely the nature of the scrapie agent, but to denigrate Stan Prusiner.
>
> I contacted Stan Prusiner some four years ago to initiate a collaboration on the scrapie problem, and this collaboration was formalized in 1985. In those four years I have come to appreciate Stan as an honest, enthusiastic and loyal collaborator, despite the fact that our opinions differ in several regards. I consider his work on the isolation and characterization of the PrP protein highly meritorious, even if its role in scrapie infectivity is not yet settled. With all due regard for the excellent work of many other researchers, I believe that many of the facts Stan has uncovered—not only isolating PrP but also making the distinction between two forms of the protein and discovering the relationship of the PrP gene to incubation-time genes—would not have been recognized so soon without his efforts.
>
> Stan's achievements were made possible by both his scientific initiative and his success in obtaining the financial support required for the very expensive animal work underlying the scrapie project. I agree with the view expressed in your article that Stan's coining of the word prion unleashed much ill feeling in the scrapie community (sheep excluded), but on whom does that reflect badly—Stan or his critics?
>
> However one may feel about the idea of a nucleic-acid-free scrapie agent, it must be clearly understood that substantial, albeit not definitive, attempts by several groups (including our own) have to date failed to reveal a defined nucleic acid in purified preparation of the scrapie agent. Those who believe that a nucleic acid—be it a "virino" or an as yet undiscovered viral particle with a high specific infectivity—is essential for scrapie infectivity carry the burden of demonstrating its existence. Conversely, as pointed out in your article, those claiming that PrP alone is the infectious agent need to demonstrate that PrP, synthesized in *E. coli* or in vitro, and possibly modified post-translationally, can cause scrapie in animals.

Stan Prusiner has an extraordinary and colorful personality, and the enthusiasm that has enabled him to carry out the back-breaking research of the past decade has led him to espouse views prematurely. However, one should remember that Howard Temin was vituperated and ridiculed by some of his colleagues for years while he pursued his vision of retrotranscription in RNA tumor virus replication, an idea that to many of us seemed every bit as outlandish then as that of an infectious protein does now. But Temin was eventually vindicated.[6]

The Taubes episode prompted me to adopt a policy of not talking to the press. It seemed to me that most science writers were happy to listen to the scrapie-virus hunters, who did not need to explain new concepts; everyone had grown up with the word "virus." Even if the details of what makes an infectious particle a virus are vague to most people, the term is familiar. The virus hunters needed only to invoke it and ask, "Isn't it obvious? Viruses cause scrapie, kuru, and CJD. Everyone knows that someday these viruses will be found." It became a hopeless battle, and I soon realized that no amount of speaking to reporters could ensure a favorable article—in fact, the opposite was true. The more I spoke to reporters, the more likely it was that an unfavorable article would appear since there was never any difficulty getting three or four scrapie-virus hunters to rebut my comments.

The press, I knew, had little influence on the outcome of scientific investigations. The opinions of neither reporters nor scientists could alter the truth. In a democratic society, each person's opinion about political issues is equally important. If out of a population of 100,000 there are 50,001 votes for, say, sodium-vapor streetlights, then the city will be obliged to replace the conventional streetlights with sodium-vapor streetlights. But in science, issues are not decided by majority vote; instead, they are resolved by careful study if sufficient technology exists to permit systematic investigation. In science, conjecture and speculation are of little value unless they suggest testable hypotheses that lead to an increased understanding of the problem. If 99,999 people vote that the earth is flat or the sun orbits the earth, the one dissenting vote will win when scientific investigations uncover the truth. Thus, I concluded that the only way for me to win this battle was to produce data and let the published record in refereed scientific journals speak for itself. And realistically, I had no other choice.

The *Discover* article's effects lingered on. In June 1987, I attended a small meeting held at the Ciba Foundation in London. The organizer, Gregory Bock, thought that scrapie was caused by a virus but had diplomatically titled the meeting "Novel Infectious Agents and the Central Nervous System." On the afternoon preceding the meeting, as I was inquiring about a room at the Ciba Foundation Building, Fred Brown, the conference chair, came up to me and said, "I've wanted to meet you for several years—you're quite an operator!" The specter of "carpet salesman" rose before my eyes, and I immediately decided this would be just another unpleasant U.K. gathering. But the next morning, after two hours of presentations and discussion, Bock cornered me at the break and admitted, "I now realize that viruses don't cause scrapie, kuru, and CJD."

"I'm delighted!" I responded.

"I'd like you to state your disagreements with Alan Dickinson during each discussion period, which is a critical part of the meeting," Bock went on. He was concerned about these discussions because they were being recorded and would be transcribed and published along with the papers submitted by the authors. He told me that he would tell chairman Brown to call on me at the end of each discussion section. By the late afternoon, I had disagreed with Alan Dickinson at least a dozen times.

Dickinson finally had enough of Fred Brown's giving me the last word. "Many of Stan Prusiner's statements are not believable," he bellowed, "and his character is suspect, as discussed at length in a recent article in *Discover*." With that, Bock jumped from his chair and took the microphone from Brown. "Another remark like that one concerning the character of any participant," he announced, "will prompt the person making the accusation to be thrown out of the meeting."

The atmosphere was as tense as it had been in earlier meetings, but now the wind had shifted—thanks, perhaps, to the *Discover* attack and certainly to an escalating accumulation of data in favor of an infectious protein. Unfortunately, I was not sufficiently secure or generous to be able to reach out to Dickinson. He must have been hurting inside. His work was being reinterpreted in terms of the prion concept; that had to be painful. He had been so sure that the scrapie agent contained a nucleic acid.

Brown's "operator" comment still rankled, and on the afternoon of the second day, I seized an opportunity to have it out with him. We went for a walk in Portland Square, strolling past old Victorian homes, many converted into offices without modifications to their elegant brick and limestone exteriors, and into the square, which was dotted with oak trees and crisscrossed with asphalt pathways. An abundance of manicured, lush flowerbeds lined these paths, a refreshing contrast from the hectic meeting room where I had been cooped up for much of the last two days. I thanked him for having been "both fair and nice to me throughout the meeting."

"Why shouldn't I be?" he said, puzzled.

"Well, you called me an operator when we met."

"So what's wrong with that?"

"When we Americans speak of people as operators, we mean they're manipulative," I said. "It's not exactly a compliment."

"That's not true in England," he informed me. "To the contrary. I was referring to how skillfully you handled yourself when Dickinson and Patricia Merz attacked you at the Sendai meeting three years ago. You were the chair of the session, and your behavior was admirable." Of course, I was delighted to hear this, but I remain bemused by the slippery differences between British and American idiomatic English.

Shortly after the meeting, as it happened, Alan Dickinson was fired from his job as director of the Neuropathogenesis Unit at Edinburgh University. The scuttlebutt was that for years members of the British scientific establishment had been frustrated by Dickinson's bravado, having repeatedly told him that he needed to isolate the gene responsible for control of the scrapie incubation time in mice. His invariable reply was, "You don't understand the experimental system of scrapie in mice." He would argue that he alone comprehended the experimental details of all his inbred lines, inoculated with different strains of the scrapie agent. An astute observer of this conflict might easily come to only two possible conclusions: Dickinson was either a genius or a stubborn eccentric unversed in molecular genetics. Apparently our latest *Cell* paper with Carlson had prompted the British funding authorities to conclude that the latter was the case and provided them with the ammunition they needed to terminate his employment. His successor was Richard Kimberlin, who resigned six months later to become a consultant to the government just as the mad cow epidemic was emerging in Britain. Kimberlin confessed to me on several occasions that he felt unqualified to direct the unit, now that so many new technologies were coming to bear on the study of scrapie.

Dickinson left a legacy that would last into the next century. His disciples were so well indoctrinated that for many years after his departure they continued at Edinburgh the so-called strain typing—which uses several lines of inbred mice to measure incubation times of different strains of prions—denying the power of transgenic, or genetically engineered, mice, which would revolutionize research on scrapie and other prion diseases. While deserving much credit for his important early contributions, Dickinson had thwarted my efforts and those of others at every turn—yet I was sad about his departure. His eccentricities would be missed.

12

Deciphering Human Prion Diseases

O ver the next two years, our discoveries would create a new paradigm of the human prion diseases that represented a major leap forward. From our data and those of others, we developed a unique understanding of how prion disorders in people can be manifest as sporadic, infectious, and genetic illnesses. Gene sequencing, transgenic mice, genetic linkage, and epidemiology would all conspire to help unravel the conundrum of human prion diseases. These studies proved to be a wonderful example of the interplay between the basic and clinical sciences—biomedical science at its best, using patients to disentangle complex biological phenomena.

The genetics of human prion disease is a story that illustrates just how poorly scientists generally see beyond their data. It begins in 1921, with the description of CJD. Most CJD cases occur in people with no family history of the disease. But by 1930, the first familial cases had been reported—in the Backer family, of Flemish descent, living in northern Germany. Paul Backer died in 1922 at age forty-four, and his sister Bertha died in 1926 at forty-six, both of a slowly progressive neurological disease. Their mother had died at age fifty-four of a neurological illness, as did seven of her nine brothers and sisters. Paul Backer's wife noted her husband's increasing forgetfulness, loss of interest, and general change in personality six months before he was hospitalized for progressive memory loss, disorientation, aphasia, and difficulty performing complex movements. He died eight months after hospitalization. Microscopic examination of his brain sections showed vacuolation and widespread loss of nerve cells, leading to the diagnosis of CJD. His sister died seven months after presenting with memory loss, ataxia, and rigidity. At autopsy, her brain showed widespread atrophy and spongiform change, prompting the same diagnosis. Paul's son Max, age forty-four, died after a seven-month progressive illness that began with impaired memory and tremor. Another son died at forty-two of a similar progressive neurological disorder, as did one of Paul's four daughters, at age forty.

Then in 1973, Raymond Roos, working with Gajdusek and Gibbs, reported transmission of familial CJD to a chimpanzee thirteen months after inoculation of an extract prepared from the brain of a man identified as R.C.[1] At age thirty-five, R.C. had died three months after the onset of progressive dementia, myoclonus, and difficulty moving his arms and legs. Sections of his brain showed loss of neurons and widespread spongiform change. His mother had died at age thirty-seven after a seven-month-long illness manifested by dementia and problems with coordination; his maternal grandmother had died at forty-eight of a neurological disease resembling her daughter's, and two maternal aunts had died of rapidly progressive dementia lasting only a few months.

R.C.'s familial form of CJD was indistinguishable, clinically and neuropathologically, from nonfamilial CJD, and, more important, like nonfamilial CJD it could be transmitted. The question of how a disease could be both transmissible and inherited was left to explanations that included the spread of the CJD "virus" among family members living in close proximity. An alternative explanation was infection at birth, a means of transmission clearly illustrated a decade later by passage of HIV from infected mothers to their babies during pregnancy, labor, and delivery, or breastfeeding. A third explanation was an inherited predisposition to infection by a ubiquitous CJD virus (that is, one occurring worldwide and randomly). But there was no evidence to support any of these hypothetical etiologies, and needless to say, no data to argue that a virus was the cause.

In 1981, Colin Masters, also working with Gibbs and Gajdusek, reported the transmission of Gerstmann-Sträussler-Scheinker (GSS) syndrome, a rare familial and invariably fatal disease with abundant amyloid plaques, to monkeys.[2] As is true of CJD victims, some GSS patients exhibit difficulty walking and, during the initial phase of the disease, are mistakenly diagnosed as having multiple sclerosis. The transmission of GSS supported the earlier study of familial CJD, and the authors concluded that "the CJD virus" caused GSS, but neither study advanced my understanding of the disease process. Only after we had sequenced the scrapie protein and retrieved cloned PrP genes could I begin to think about investigating inherited forms of these two diseases.

GSS was first described in 1936, by Josef Gerstmann, Ernst Sträussler, and Isaak Scheinker, all of whom were Jewish Viennese neuropsychiatrists.[3] In 1928, Gerstmann reported a woman, Berta H., age twenty-six, with difficulties in maintaining her balance, which progressively worsened. He also described her change in mood and character: A cheerful woman, she remained so, but she also became irritable, intolerant, and apathetic. She had problems with coordination, both in moving her arms and legs and in speaking. Her reflexes were absent, although she had a positive Babinski reflex—where the big toe goes upward when the sole of the foot is scraped with a pointed object—in both feet. Damage in the cerebral hemispheres often results in a positive Babinski reflex. Berta H. died six years after the onset of her illness. Her father, grandfather, and great-grandfather had all died of a similar disease, their illnesses beginning at ages forty, thirty, and forty, respectively. Further investigations in the early 1990s of the family showed that the great-grandfather's sister had also died of this illness, as had two of her sons and one grandson. The family now encompasses more than 221 members in nine generations, in which 20 definitely had GSS. Another 16 may have had GSS, and 12 had neurological and psychiatric diseases said to be unrelated to GSS. At autopsy, the brain of one of these patients, H.M., showed extreme atrophy, and microscopic examination of brain sections revealed severe loss of neurons, widespread spongiform change, and numerous amyloid plaques.[4]

My own clinical experience with GSS began courtesy of Bruce Miller, who was then at UCLA (figure 26). Knowing of my interest in CJD and scrapie, he described a case to me: "He's a forty-one-year-old white male who began having difficulty walking about two years ago. He has limb and truncal ataxia, but he's not demented. His older sister had a similar presentation about ten years ago, was initially diagnosed with multiple sclerosis and later with olivopontocellular atrophy, but at autopsy the diagnosis of GSS was made by a neuropathologist at UCLA."[5]

"So that's why you think her brother also has GSS?" I asked.

Figure 26   In 1987, Bruce Miller (left) showed me a patient at UCLA who was dying of GSS. Two years later, Karen Hsiao (right), along with Tim Crow, Rosalind Ridley, Harry Baker and Jurg Ott, reported that a mutation in the patient's PrP gene was genetically linked to the development of GSS.

"I'll let you know the next time I'm planning to see him in my clinic."

"That would be great!" I said. "We need to obtain some blood so we can isolate the DNA from his white blood cells. I want to find out whether he carries a mutation in the DNA of his PrP gene that results in a change in the amino-acid sequence of his prion protein."

We isolated the DNA from Bruce Miller's GSS patient and stored it in a freezer, until seven months later, when Karen Hsiao (see figure 26) came to work in my laboratory. Later, Bruce Miller and I would become close colleagues after I recruited him to head our clinical dementia program, now called UCSF's Memory & Aging Center. Bruce's remarkable talent for organizing clinical research programs led to the creation of a world-renowned program for the study of the frontotemporal dementias. After completing her neurology residency at UCSF, Karen wanted to study familial human prion diseases. Though Karen had been an excellent neurology resident, she brought an intensity and focus to her laboratory work that was rarely matched. She was already well schooled in the art of scientific investigation through her neuroanatomy studies that had earned her a Ph.D. Her project was to clone the PrP gene from Miller's patient and sequence at least ten different clones, each containing the entire PrP gene sequence. Since GSS is dominantly inherited, only one of a patient's two PrP genes needs to carry the fatal mutation. Today such a study can be done quickly, using a technique known as PCR (for polymerase chain reaction), but at that time sequencing multiple clones was the only way to find a mutation in a dominantly inherited disease.

Eventually, Karen found a mutation, a single base change in the DNA sequence: At codon 102, the DNA sequence CCG was mutated to CTG, resulting in the substitution of the amino acid leucine for proline.[6] We immediately surmised that this single base change was responsible for the degeneration of the nervous system of the patient as well as that of his sister. But I argued for caution: First, we had to be sure that this mutation was not found in the general population; in other words, was it specific to families with GSS? Then, once the mutation was associated with the disease, we would have to establish its statistical significance by a genetic-linkage study before publishing our findings.

To perform a genetic-linkage study, we needed more people with GSS—particularly people from large families. Tim Crow and his colleagues, working at the Institute of Psychiatry in London, had assembled a large pedigree of the W. family, of which twenty-three members over seven generations were affected. The disease generally began insidiously; often, the patients were unsteady and forgetful. Like patients with kuru, their difficulties in standing worsened with time, due to truncal ataxia, and they had problems with coordinated movements of the arms and legs. Most of the affected family members developed a progressive dementia. The length of the clinical disease varied from one to ten years, but four years was typical; at autopsy, the patients' brains showed widespread spongiform change and numerous amyloid plaques. I asked Crow to collaborate with us, and soon we discovered that the British GSS patients all carried the same mutation that Karen had found in Bruce Miller's patient.

I next called Jurg Ott, a professor of statistical genetics at Columbia University, and asked him to help us assess whether we had a sufficient number of individuals to determine whether the mutation was significant. Jurg did a series of calculations and told us that we did not have sufficient data but that with DNA samples from about a dozen healthy young family members we could infer who among the dead parents had carried the mutation. Fortunately, Tim and his colleagues had built an excellent relationship with the W. family, and they were able to obtain the needed samples. Once we had a linkage score showing that the odds were more than 1,000 to 1 that the mutation we had found was the cause of GSS, we hastily wrote a paper and sent it to *Nature*.[7]

While the paper was being reviewed in the fall of 1988, I was in Tokyo for a prion meeting and reconnected with my friend Jun Tateishi. I'd met Jun in Fukuoka, Japan, in 1978, on my way back from my first trip to the highlands of New Guinea. Jun was a thin man of average height with prematurely gray hair; his personality was that of a reserved, soft-spoken, kind, and gentle person who was always proper and appropriate. He had transmitted a brain extract from a patient who died of GSS to mice, and extracts from these mouse brains transmitted the disease to inoculated mice in about 150 days. I liked the idea of using mice to study the human prion diseases; they're easier to deal with than monkeys. Jun sent me a tube containing a brain from a mouse inoculated with the Fukuoka strain of GSS. I inoculated it into a dozen mice and waited about 150 days until they became ill. When I announced to my coworkers that the brains of these mice were available for study, there was an uproar I never anticipated. Several postdocs and technicians said they would quit if I began studies with a human pathogen. I also learned that several other faculty members with laboratories nearby were upset. With help from one of the technicians, I came in on a Saturday, sacrificed these animals, stored each one in its own plastic tube, and immersed them all in liquid nitrogen. Darlene Groth would periodically ask me, "Stan, can we throw out those GSS mice? We need more freezer space." After a decade, I finally capitulated and told her to go ahead and incinerate them.

At the end of the Tokyo prion meeting in 1988, I sat down with Jun in the hotel coffee shop, told him about Karen's discovery of the mutation that caused GSS, and swore him to secrecy. "What I would like from you," I said, "are blood samples from Japanese CJD patients where there is a family history of CJD." Jun promised to send some, but three weeks later what I got instead was a manuscript describing a hastily performed GSS study of his own, showing the same mutation in the PrP gene in eleven Japanese people with GSS and another mutation, at a different position in the gene, in a French GSS patient. He suggested publishing it back-to-back with our paper in *Nature*. I was unhappy. I called him and told him that Karen Hsiao had worked on the GSS study for almost two years and should be given the credit she deserved. A few days later, Jun visited me in San Francisco. I was still upset, and when he walked in the door I said, "Jun, you cannot do this. It's not right. We worked for two years on this project. You spent a few days, after I asked you for samples from patients with a family history of prion disease."

"But we just want to publish our paper with yours," he said. "It can only make your paper better."

I was getting heated. "Jun, if we had *not* done a genetic-linkage study showing that the mutation causes GSS, and you *had* done one, then you would have a point. But the opposite is the case. We did all the work, and all you've done is report the presence of mutations, which is meaningless until the linkage study has been done."

Sometimes the competitive nature of science pushes investigators, including myself, to be inappropriately aggressive. To Jun's credit, he quickly grasped my argument, and though disappointed, delayed publication of his paper by more than six months. At the end of this somewhat tense exchange, he peered at me softly and added, "All these English words are giving me a headache."

Three months later, with our paper due to appear soon in *Nature*, I had still not learned my lesson. I was at a meeting in Washington, D.C., having dinner with one of Carleton Gajdusek's postdocs and an NIH grants administrator, my head still whirring with the excitement of our molecular genetic findings. During dinner, the use of molecular genetics to study Alzheimer's disease came up, and I could no longer contain myself. I proceeded to explain our results with GSS. My dinner companions listened intently.

Within a week, Gajdusek and his postdoc had submitted a paper to *Nature* reporting the mutation at codon 102 but without a linkage study. I was later told by one of the authors that *Nature* returned their paper without reviewing it. But our discovery of pathogenic mutations of the PrP gene in familial prion diseases had given Gajdusek a new lease on his scientific life. He'd been collecting samples of CJD brains for nearly twenty-five years. The samples were stored in numerous freezers, and now all his coworkers had to do was isolate the DNA and sequence the PrP gene. As he began to report new mutations, I started to worry that he might obscure our discovery. This never happened, because he didn't understand the meaning of these mutations: He thought they might confer susceptibility to a ubiquitous infectious agent in the environment—namely, the elusive scrapie virus. The time we took to establish genetic linkage between the mutation in the PrP gene and the development of GSS had been close to disastrous for us. DNA sequencing technology had improved so much in the meantime that others could replicate in a few days what had taken us many months to do. Subsequently, more than forty different mutations have been found to cause inherited forms of the human prion diseases.

Demonstrating significant linkage between the PrP gene mutation and the neurological disease GSS altered my thinking in ways I did not anticipate. I never dreamed that our finding of a PrP gene mutation would prove to be such a powerful argument against the putative scrapie virus. Our work clearly showed that GSS and familial CJD were genetic diseases caused by mutations in the PrP gene. When people died of inherited prion disease, their brains contained infectious prions. The most likely explanation was that mutant PrP^C was much more readily converted into the disease-causing PrP^{Sc} than was normal PrP^C. Later, we and other investigators showed that if people carrying PrP mutations live long enough, they all develop prion disease.

We had discovered an entirely new mechanism of disease. We had shown that a single illness can be both genetic and infectious. We could also explain a third manifestation of human prion disease: the "sporadic" form of CJD. More than 85 percent of all CJD cases are sporadic CJD—that is, there is no history of exposure to prions through eating or other routes, and no family history. Sporadic CJD occurs mostly in people between fifty and seventy-five years old; in people older than that, it is sometimes mistaken for Alzheimer's disease. We posited that sporadic CJD results from a rare (one in a million) spontaneous, self-sustaining conversion of PrP^C into PrP^{Sc}. The spontaneous formation of PrP^{Sc} in people with sporadic CJD explained, for one thing, why such patients are found all over the planet: The incidence of CJD is the same no matter where you look.

Interlude: Three Hypotheses That Might Explain the Spontaneous Formation of PrP^{Sc} Prions

Before describing three hypotheses, one or more of which might explain the sporadic form of CJD, I must introduce a few terms. When a mutation within the coding region of a gene occurs, there is a change in one of the bases of the DNA. For example, when the codon "CCA" encoding the amino acid proline is changed to "CTA," then the amino acid leucine is substituted for proline. However, when "CCA" is changed to "CCG," there is no change in the amino acid sequence due to the degeneracy of the genetic code. Mutations in DNA that do not lead to changes in the amino acid sequence of the protein are called silent.

Besides coding and silent mutations, there are germline and somatic mutations. Germline mutations occur before conception and are passed onto children through the sperm or the egg. In contrast, somatic mutations occur after conception in any cells of the body except the sperm or egg, and thus are not passed onto children.

With the results from our studies on inherited forms of prion disease, in which a PrP gene mutation is found in the germline cells—sperm or egg—we offered three hypotheses, any of which might explain sporadic CJD, which accounts for more than 80 percent of all cases:

(1) We proposed that PrP gene mutations in somatic, nongermline cells might stimulate conversion of PrP^C into PrP^{Sc}. Once PrP^{Sc} accumulated due to a somatic mutation, it might stimulate conversion of nonmutant, wild-type PrP^C into more PrP^{Sc} molecules.

(2) We suggested that the flipping of PrP^C into PrP^{Sc} was a chance event that, once in a great while, occurred because of some undefined metabolic condition, such as the accumulation of a toxic metabolite.

(3) We also raised the possibility that the formation of low levels of PrP^{Sc} might be a normal process and that PrP^{Sc} might have a physiological function. As long as cells, particularly neurons in the brain, were able to degrade PrP^{Sc} as rapidly as it was formed, everything would be all right, but the moment PrP^{Sc} began to accumulate faster than it could be cleared, the process embarked down a fatal pathway.

Slowly, opposition to the prion concept was collapsing. I was particularly happy to get a call from Joe Gibbs in the fall of 1988, asking me to have lunch with him on my next trip to Washington. We met at the Officer's Club at the Bethesda Naval Medical Center. Gibbs began by saying, "My four young scientists were charged with showing that your work is wrong. Instead, they have reproduced every aspect of your studies."

"That's great news. When are you going to publish your findings?"

"Soon. We've completed so many studies that we have enough data for four manuscripts. Can you help us find a place to publish?" Since his group had simply confirmed my work rather than broken new ground,

Gibbs was concerned that no journal would be interested in publishing the studies.

"I'll try," I said, possibly more eager than he was to see this important confirmatory work in print. A few days later, I lobbied the editor of *Neurology* to look kindly on the studies, and they were eventually published in that journal.[8]

Nonetheless, the scrapie-virus hunters remained undaunted. Soon after our publication of the PrP mutation and the genetic-linkage study, some of them, including Richard Kimberlin and Bruce Chesebro, began to argue that PrP is a receptor for a ubiquitous virus. They contended that mutant PrP binds the "scrapie virus" with increased affinity, which results in more efficient infection. However, by now many strains of prions had been identified, and the prions of different mammalian species exhibited different host ranges. As the tally of different prions kept expanding, the difficulty of explaining how inherited prion diseases are caused by viruses was becoming overwhelming. The scrapie-virus proponents had to postulate a large array of ubiquitous scrapie viruses, consisting of hundreds, if not thousands, of varieties—yet no one had found even a single scrapie virus, anywhere! The only way out of this predicament for the scrapie-virus hunters was piling on one ad hoc hypothesis after another. At times, I almost felt sorry for them.

Our work was still descriptive. We had not yet managed to penetrate the mechanism of scrapie-prion multiplication. With Michael R. Scott, who had joined my group in 1986 after extensive training in molecular biology with Harold Varmus, I decided that the next step would be to construct mice expressing a hamster PrP gene—or transgene, as it is commonly called (a contraction of "transferred gene") (figure 27). Mice that carry transgenes are called transgenic, or Tg, mice. We knew, from our own work and that of others, that infectious prions did not readily pass from mice to hamsters and vice versa. If we could abolish this transmission barrier by expressing hamster PrP in mice, we might substantially advance our understanding of prions. Moreover, once such Tg mice had been inoculated with hamster PrP^{Sc}, we would also be able to test whatever prions were produced in their brains in nontransgenic mice and hamsters.

The barrier for transmission of prions between hamsters and mice was reminiscent of the "species barrier" described by Iain Pattison in the

Figure 27   Mike Scott constructed transgenic mice expressing hamster PrP^C.

early 1960s. Working at the animal-disease research institute in Compton, U.K., Pattison had studied the passage of the scrapie agent between sheep and rats and later between rats and mice. He found that on first passage of the agent from one species to another, the incubation time was prolonged. On the second passage within the same species it shortened, and it remained constant in subsequent passages. The commonly held interpretation of these findings on the transmission of scrapie from one species to another was that the "scrapie virus" needed to adapt. When the putative virus was transmitted from one species to another, a few of the "viruses" were thought to mutate and then multiply. Since adaptation of viruses from one species to another is well documented and known to be due to mutations in the viral genome, this was offered as a compelling argument for the existence of the scrapie virus.

Mike and I thought that one way of testing this adaptation argument was to construct Tg mice by inserting the hamster PrP gene in their genome and then to inoculate them with hamster PrP^{Sc} and see how long it took them to develop the symptoms of scrapie. If they did so within the normal time frame for onset of the disease, then a virus was probably not involved.

Constructing such mice proved to be complicated. I tried to get an expert at the University of Pennsylvania to make the Tg mice for us, but he was inundated with requests for help with projects deemed to have a much higher chance of success. Mike engineered a minigene by linking together small pieces of hamster DNA, but we soon found that this approach was not useful; minigenes had been reported by others to fail often in Tg mice. So Mike undertook the job of isolating a large piece of hamster DNA that contained the entire hamster PrP gene and much more. Setting up a facility to microinject Mike's large piece of hamster DNA into the fertilized eggs of mice was also a challenge. We needed an isolated space, so that the surrogate mothers, in whom we had implanted the eggs containing hamster DNA, would not miscarry. Moreover, the space needed to be close to our laboratory so I could oversee the project. Fortunately, the chairman of orthopedic surgery was willing to lend me the space, but there was a price to pay. I had to listen to his opinion of neurologists, which was low. "You'll never cure CJD," he said. I bit my tongue. "I really appreciate this," I said. "Few chairs would be so kind."

By 1989, after almost two years, we finally had Tg mice whose genomes contained hamster PrP. Would they develop scrapie rapidly after inoculation with hamster scrapie prions, or would they show prolonged incubation times, as do nontransgenic mice that are inoculated with hamster PrP^{Sc}? Could the hamster PrP gene, by itself, abrogate the transmission barrier?

Darlene Groth phoned me with some exciting results from our experiments while I was on a trip in the beginning of summer of 1989. Returning late on a Friday, I decided to wait until Saturday morning to examine the Tg mice. I drove to the Berkeley vivarium, where our mice, now safely beyond the ken of the chairman of orthopedic surgery, were housed. After traversing an elaborate security system, I found the cages with the Tg mice Darlene had described. These mice had been inoculated only seventy days earlier with hamster scrapie prions, and they were all showing neurological signs of scrapie. The results were stunning! The hamster PrP transgene not only rendered the mice susceptible to hamster scrapie prions, but it also enabled neurological dysfunction to occur in half the time required for mouse prions to do so. Instead of a few wild-type mice developing scrapie more than five hundred days after the time of inoculation, all of our Tg mice were developing scrapie in a little over two months.

At this point, I knew we had finally cracked the scrapie problem. The results argued that a virus did not cause scrapie and that prions were real. More important, in transgenic mice we now had a tool that allowed us to explore in ways no one had investigated previously. We could do thousands of experiments to define virtually every aspect of prion biology. No longer would all of our experiments be descriptive. For the first time, we could address mechanistic issues, simply by changing the sequence of the PrP transgenes.

Driving back across the Bay Bridge, I contemplated all the wonderful experiments we could now undertake: We were poised to study prions from different species, including humans with CJD, using transgenic mice. An entirely new world of prion biology had just opened up. Only insufficient funding limited the speed of the torrent of new information about prions. As I drove, too few of my neurons were focused on the traffic, and all of a sudden the car in front of me stopped. I slammed on the brakes, but not soon enough. Fortunately, no one was hurt, and remarkably the car I had plowed into was undamaged. I drove on, a chastened biologist.

As multiple lines of Tg mice expressing hamster PrP became ill after inoculation with hamster PrP^{Sc}, we had the information that Mike Scott needed to write a paper.[9] We found that the brains of our inoculated Tg mice contained a billion infectious hamster prions and, as expected, lots of hamster PrP 27–30. What surprised Steve DeArmond and me were the large amyloid plaques in the brains of these Tg mice. Our findings contended that the PrP gene modulates the susceptibility, incubation times, and neuropathology of prion disease and demonstrated that infectious prions could be programmed by a genetically engineered PrP transgene. Although many investigators greeted our work enthusiastically, the scrapie-virus hunters expressed their usual skepticism. Just as they had struggled to explain inherited prion diseases, they insisted that PrP was a receptor for the "scrapie virus" and that hamster PrP rendered mice susceptible to it. But the data that emerged over the following year from additional experiments with our Tg mice were even more difficult for the virus hunters to explain.

Our Tg mice expressed both mouse and hamster PrP. The mouse PrP was encoded by the naturally occurring chromosomal PrP gene that all mice have, whereas the hamster PrP was encoded by the transgene we had inserted into the mouse embryos. So we could now determine what kinds of prions would assemble in the brains of our Tg mice in response to inoculation with mouse scrapie prions or hamster scrapie prions. Would they produce a collection of mouse and hamster scrapie prions? Or would they produce a hybrid prion, and would it be equally effective in attacking hamsters and mice? Tg mice were sacrificed when they developed signs of scrapie, and their brains were removed for inoculation into hamsters and nontransgenic mice. Our experiments to determine what types of scrapie prions were produced in the Tg mice required more than a year to complete.

The striking results told us that mouse prions selectively stimulate the conversion of mouse PrP^C, the benign form of the prion protein, into mouse prions composed of mouse PrP^{Sc}. Conversely, hamster prions stimulate the conversion of hamster PrP^C into hamster prions comprising hamster PrP^{Sc}. Our results were so clear-cut that we wondered whether we had made a mistake. Out of 231 amino acids comprising the prion protein, only 11 differ between mouse and Syrian hamster—one or more of these 11 had to control whether mouse or hamster prions were produced in our Tg mice.[10] The fundamental process of scrapie-prion multiplication had been revealed by what now seemed to be a rather simplistic set of experiments—an unprecedented process, occurring in the absence of a nucleic acid.

Steve DeArmond's examination of the Tg mouse brains was equally remarkable: He discovered new principles governing amyloid-plaque deposition. We learned that the strain of inoculated prion—mouse or hamster—determined whether PrP^{Sc} was deposited as amyloid plaques in the brain. When the Tg mice were inoculated with hamster prions, they developed neuropathologic changes similar to those seen in scrapie-infected hamsters, including numerous PrP-amyloid plaques.[11] When the Tg mice were inoculated with mouse prions, they developed neuropathologic changes similar to those seen in nontransgenic, scrapie-infected mice, where PrP-amyloid plaques were rare—showing that amyloid plaque formation was determined by the strain of prion in our Tg mice. The type of prion, and not the host, determined whether amyloid was deposited.

After a long period of stability in the housing of our hamsters and mice, we were rudely told one day the UC Berkeley campus had written us out of their future vivarium plans. They decided that our exit would expedite their quest for State of California funding to build a modern, centralized vivarium. With our departure, they could show that their new facility would be smaller than the sum of all their current small ones and thus garner modest support from groups opposed to animal experimentation.

Fortunately, Julie Krevans and his executive vice chancellor David Ramsay found a superb solution to my newest animal crisis. They generously made available a facility at the old Hunter's Point Naval Yard, which had been used for laboratory animal studies in World War II. For the first time, we had adequate space to build two separate colonies of Tg mice: one for scrapie studies and one for the investigation of human prion diseases. David Ramsay's exceptional vision on where we might be able to take our work made all the difference.

Interlude: Transgenic Mouse Studies of GSS

Because inheriting a PrP gene mutation from only one parent is sufficient to cause GSS, it should be possible to create a model of GSS in mice—that is, Tg mice expressing normal as well as mutant PrP should become ill, since this is the case with people who develop GSS. Karen Hsiao followed Mike Scott's lead and inserted the human leucine mutation from GSS patients into mouse PrP.* And indeed, by three months of age, some of these mice had developed symptoms of a neurological disease. When Steve DeArmond examined their brains, he found changes characteristic of the GSS form of prion disease: spongiform appearance and numerous prion-amyloid plaques. We had taken the mutation causing GSS in humans and re-created it as a single base change in the mouse PrP gene. When the mutant mouse PrP transgene was expressed in mice, it produced neurologic disease indistinguishable from that caused by inoculation with prions. This finding was another compelling argument that inherited human prion diseases were caused by PrP gene mutations, not by a ubiquitous scrapie virus lurking all over the planet.

*K. K. Hsiao, M. Scott, D. Foster, D. F. Groth, S. J. DeArmond, and S. B. Prusiner, "Spontaneous neurodegeneration in transgenic mice with mutant prion protein," *Science* 250: 1587–90 (1990).

But again the scrapie-virus hunters rose up, contending that mutant PrP was a receptor that allowed a ubiquitous virus to infect our Tg mice.

Among them was my collaborator Charles Weissmann. At a private annual review meeting in San Francisco in 1991, Weissmann described his idea of a small nucleic acid attached to the prion. He was reinventing Alan Dickinson's virino from the 1970s, but he was too clever to call it a virino—he decided on "holoprion." The holoprion was composed of the apoprion, which was PrP^{Sc}, and the coprion, which was a small RNA. When the coprion RNA was bound to PrP^{Sc}, it could specify properties of the various prion strains; in the absence of the coprion, the apoprion was still infectious. The holoprion hypothesis was published in *Nature*,[12] and Weissmann began pushing this idea hard on the lecture circuit. The ingenuity of the virus hunters was frustrating on one hand and admirable on the other. They seemed to counter every experimental study with another ad hoc argument. They had no consistent set of objections, only a never-ending stream of quibbles, no matter how convincing our data were.

By late 1990, the list of PrP gene mutations in GSS and familial CJD in people had grown to more than a dozen. In the fall of 1991, Pierlugi Gambetti, director of neuropathology at Case Western Reserve's medical school in Cleveland, called and asked for some of our PrP antibodies. Gambetti was working with Elio Lugaresi, chairman of the neurology department at the University of Bologna, who had described several Italian families six years earlier with a disease called fatal familial insomnia, or FFI. The patients suffer from progressive insomnia and difficulties with body temperature regulation, urination, and bowel movements. Eventually, many develop slurred speech, tremors, myoclonus, and severe fatigue. Typically, these patients die about a year after the onset of insomnia. At autopsy, the changes in the brain are minimal, but there is a loss of nerve cells in parts of the thalamus, one of the major deep structures of the brain, on which the cerebral hemispheres are perched. The spongiform degeneration that Gambetti occasionally saw in the thalamus made him wonder if there were deposits of PrP^{Sc}. When he used our antibodies, he found intense staining in some regions of the thalamus. After he and Lugaresi were convinced that PrP^{Sc} was deposited in the brains of patients with fatal familial insomnia, they used polymerase chain reaction (or PCR, as it is often called) and rapidly found a mutation at codon 178 in the PrP gene of these patients. When their study was published, it widened the spectrum of prion disease.[13]

The invention of PCR had simplified DNA sequencing, and the hunt for mutations in dominantly inherited diseases intensified wherever the responsible gene had been identified. Neurologists and neuropathologists all over the world were studying more than a hundred families with familial prion diseases. For example, in 1990, Frank Owen and John Collinge, using PCR and working with Tim Crow, demonstrated that the change in codon 129 from ATG to GTG is a polymorphism and not a mutation.[14] A gene polymorphism describes a variant in the general population, whereas a mutation is rare. The DNA sequence ATG encodes methionine (M) and GTG encodes valine (V). By the following year, Collinge, who had moved to his own laboratory in London, reported that most people with sporadic CJD were homozygous at codon 129 for either M or V: Thus, people who are MM or VV are predisposed to sporadic CJD, whereas those who are MV are somewhat protected.

The power of mouse and human genetics for studying prions proved awesome. Tg mice expressing foreign, mutant, or artificial PrP genes propelled the science of prions forward, allowing many hypotheses to be tested. As long as the mice developed prion disease, then the results were unambiguous. For the first time, prion studies were limited only by the cleverness of the scientists who posed the questions.

13

What's in a Shape?

Just what exactly turns the harmless form of the prion protein into its evil twin? This was the great, still-unanswered question. From transgenic mice, we learned that new prions were generated when the good twin danced with the evil one. We surmised that good guys needed to come in contact with the bad ones as new prions formed. Although prion studies driven by molecular biological and genetic investigations had created a legion of supporters, the utility of that approach waned in the early 1990s, and protein structure research soon assumed a predominant role.

We had begun by searching for a chemical modification that might distinguish the prion, PrP^{Sc}, from its benign precursor protein, PrP^C, hypothesizing that such a chemical modification could be what turns PrP^C into PrP^{Sc}. For nearly six years, from 1986 to 1992, Neil Stahl, an exceptionally talented postdoc, took on this task. He cut PrP 27–30 into pieces with special enzymes called proteases and measured the mass of each piece. If a chemical group had been added to the amino acid chain during the formation of PrP^{Sc}, we should have been able to find it. But no such change showed up, and this made Neil and me nervous. Were we missing something? What we needed was an expert in mass spectrometry to measure the mass of each PrP^{Sc} fragment, to see whether its mass was due solely to its amino acid backbone or whether there were some additional atoms. When Michael Baldwin came to UCSF on a sabbatical to work with Al Burlingame, we found our man. Mike had set up the mass spectrometry facility at University College London and had the skills we lacked.

Neil and Mike searched for a way to explain the existence of two forms of the prion protein. Since the sequence of PrP 27–30 was the same as that predicted by the sequence of the PrP gene, we could not ascribe prion infectivity to a difference in the amino acid sequences between PrP 27–30 and the precursor PrP^C. Convinced that the amino acid sequences of PrP 27–30 and PrP^C were identical, I hoped that Neil and Mike employing mass spectrometry might be able to find some chemical group that had been added during the conversion of PrP^C into PrP^{Sc}. Using our most highly purified preparations of prions, they failed to identify such a chemical group. Why couldn't they? Were they overlooking something, or did it not exist? Neil and Mike were superb chemists, so I felt it was unlikely that they missed the chemical modification we all sought.

Once a protein's chain of amino acids is assembled, the protein is inactive until it folds into an active shape. Might the profoundly different properties of PrP^C and PrP^{Sc} be due to differences in their shape? For many years, the folding of a protein was thought to be dictated only by the order of its amino acids. In the early 1960s, Christian Anfinsen, working at the NIH, had discovered that the amino acid sequence of ribonuclease—or RNase, as it is often called—specifies a biologically active shape. He was able to unfold (denature) RNase and then refold it into an active conformation, whereby its RNase activity was restored. I was at the NIH in the early 1970s, working in Earl Stadtman's laboratory, and I remember the excitement surrounding Anfinsen's studies. Those studies, for which he was awarded a Nobel Prize in 1972, led to a rather dogmatic view of protein folding: that the protein's amino acid sequence determines the protein's shape. That is, each sequence produces a protein that folds into one, and only one, biologically active conformation. Prions would change this dogma.

When I told one of my colleagues at UCSF about our inability to find a chemical difference between PrP^C and PrP^{Sc}, he suggested that Fred Cohen might be helpful. Fred was trained as an endocrinologist and biophysicist; he was an expert in computational studies of protein structure (figure 28). He examined the problem and said he would be happy to help—though he had to ignore a certain amount of derision from other UCSF scientists about the possibility of ruining his career should he venture into prion research.

Figure 28   Unable to find a chemical modification that distinguished PrP^{Sc} from PrP^C, Keh-Ming Pan (left) purified PrP^C using gentle conditions so as to maximize the likelihood that the protein would retain its native conformation. Having native PrP^C was the critical preparation that we needed to be able to test a model proposed by Fred Cohen (right), which suggested that PrP^{Sc} and PrP^C have very different physical structures.

The two major structural motifs in proteins—alpha helices and beta sheets—are very different. An alpha helix is composed of amino acids that are wound in a series of right-hand turns. In contrast, beta sheets comprise flattened strands of amino acids. The sizes and numbers of alpha helices and beta sheets in different proteins vary widely.

Because of the polymerization of PrP^{Sc} into amyloid, which has high beta sheet content, Fred and I thought PrP^{Sc} was likely to have a high beta sheet content. Over the next year, three groups, including my own, measured the beta sheet content of PrP 27–30. Byron Caughey and his colleagues, working with Bruce Chesebro at the Rocky Mountain Laboratories, were the first to publish their findings, which showed that PrP 27–30 amyloid fibers were indeed just like those of many other amyloids, with high beta-sheet content.[1]

But we had no clue about the structure of the precursor protein, PrP^C, so the next task was to purify PrP^C and determine its structure. Several months before Caughey's paper was published, Fred showed me some preliminary computer models. PrP^C seemed to be folded into a structure composed largely of alpha helices and little or no beta sheet. I thought we should be able to distinguish easily the differences between the shapes of PrP^C and PrP^{Sc} if Fred's model of PrP^C was a reasonable representation of its structure. If PrP^C had a high alpha-helical content and little or no beta sheet, then the acquisition of beta sheet might be the fundamental event in prion formation.

Our minds were racing ahead. Could this puzzle turn out to be so straightforward? But much work had to be done before we learned just how accurate was our glimpse of the structural differences between PrP^C and PrP^{Sc}.

Fred found five potential sites for alpha helices in PrP^C; however, depending on how he biased his modeling programs, some of the sites could be beta sheets instead. Now we began to build models of the various shapes that PrP^C might assume after it folded. But the key experiment would require purification of PrP^C in its native shape, followed by determination of its structure. Comparing the structure of PrP^C to that of PrP^{Sc} would be meaningful only if these two forms of PrP had vastly different amounts of alpha helix and beta sheet.

Purification of PrP^C was no easy task. It is approximately ten times less abundant than PrP^{Sc} in the brain, and it is unstable. Grabbing PrP^C by using antibodies attached to beads seemed like a good idea. Antibodies would be able to find PrP^C among thousands of other proteins and bind to it. But there were two problems: First, PrP^C is a sticky protein because of a complex lipid that anchors it to the surface of a neuron, so detergents would be needed to nullify this stickiness. Such detergents make the use of antibodies complicated, since antibodies generally work best in an aqueous environment free of detergent. Second, and more problematic, was figuring out how to release PrP^C from the antibody without unfolding it, which would render structural studies meaningless. In 1990, when Keh-Ming Pan joined the laboratory as a postdoc (see figure 28), I asked him to modify the purification of PrP^C, hoping we could obtain large quantities from the brains of uninoculated hamsters. Keh-Ming managed to improve the purification scheme substantially, but he was forced to unfold the PrP^C when he released it from the antibodies. We needed PrP^C in its native, folded state, and after several false starts we were finally able to obtain it.

Interlude: Purifying PrP^C

The purification of PrP^C was complicated by the glycolipid that was attached to one end of the protein. In Neil Stahl's initial search for modifications of PrP^{Sc} that distinguish it from PrP^C, he discovered a portion of the glycolipid. The unusual chemical group that he found provided the clue we needed to discover that PrP^C is tethered to the surface of cells via a glycolipid anchor. The presence of the anchor told us that we had to use detergents to neutralize the stickiness of the lipid piece of the anchor in our purification protocol.

Over the course of nearly two years, we tried several dozen ways to purify PrP^C. Eventually, we settled on what is called immobilized metal ion affinity chromatography or IMAC because PrP^C binds copper ions tightly. But passing PrP^C over an IMAC column was still not enough to achieve the level of purity that we needed. So we turned to the sugar-binding proteins called lectins, which bound to the two sugar chains attached to PrP^C. The Swedish virologist Erik Lycke had studied the interaction of PrP 27–30 with various lectins a few years earlier, and his work pointed us to the best of the ones to use. With the PrP^C still not quite pure, we bound it to the lectin wheat germ agglutinin and washed away the unwanted proteins. To release the PrP^C from the lectin, we washed the beads with the sugar N-acetylglucosamine, under gentle conditions that would not unfold it. When Keh-Ming measured the amount of alpha-helical and beta-sheet structure in it, Fred Cohen's prediction that PrP^C was composed largely of alpha helices was verified experimentally.

Keh-Ming collected a remarkable set of data: The amino acid chain that folded into PrP^C created a protein consisting of about 50 percent alpha helix and less than 5 percent beta sheet. By contrast, PrP^{Sc} comprised about 30 percent alpha helix and almost 50 percent beta sheet. No longer did we need to think of PrP^C and PrP^{Sc} as little black boxes: For the first time, the structural change underlying the conversion of PrP^C into PrP^{Sc} was known.[2]

Besides the importance of our results for understanding scrapie and its related human diseases, there were profound implications for much of biology and medicine. For more than two decades, the dogma of protein folding from Anfinsen's work on RNase stated that the amino acid sequence of a protein determines its conformation. But we now had clear evidence for a protein that existed in two distinct conformations: In one shape (PrP^C), it performed some as-yet-unknown normal function; in a second shape (PrP^{Sc}), it became infectious and eventually killed its host.

Our finding of a protein that exists in two different biologically active conformations proved as revolutionary as the original discovery of prions. Throughout the 1960s, numerous studies showed that many enzymes could adopt different conformations, but in each case a small chemical substance, such as a sugar or amino acid, initiated the conformational change. Typically, the binding of a sugar or amino acid triggered an alteration in the shape of the enzyme that in turn modulated the catalytic activity of the enzyme either up or down. In the case of the prion, no such small molecules were involved in shifting the shape from PrP^C to PrP^{Sc}. Initially, many protein chemists and biophysicists were skeptical, but in time they greeted our discovery with enthusiasm. As techniques for monitoring protein structure improved, evidence for multiple conformational states of other proteins began to emerge. The dogma that had grown out of Anfinsen's elegant studies on the folding of RNase into an enzymatically active conformation disappeared with a whimper. The next step would be to determine the precise three-dimensional structures of PrP^C and PrP^{Sc}.

Interlude: Protein Plasticity and Alternative Folding

There is a remarkable beauty to the complexity of proteins and the simplicity of their assembly (see chapters 2 and 10). Our DNA carries instructions about how to make proteins. Although we have about twenty thousand genes, we can create a much greater number of possible proteins due to rearranging gene segments during the synthesis of messenger RNA. The brain is thought to contain more than one hundred thousand different proteins.

After chains of amino acids are assembled, they must fold into a particular shape before the protein can become biologically active. Sometimes proteins require help to fold properly—other proteins called chaperones provide this help. Cells have mechanisms to degrade improperly folded proteins. For an enzyme, which is also a protein, to be active, it must fold into a shape that allows it to catalyze or accelerate a chemical reaction. For example, each member of a group of enzymes that break down sugars must properly fold in order to perform its respective role in metabolism.

Scientists generally refer to the shape of a protein as its conformation. In the 1960s, some seminal studies demonstrated that only one conformation permitted an enzyme to be highly active. But soon other studies showed that some metabolites could modify the shape of enzymes, leading to an alteration in their activity. If the metabolite bound to a site other than the enzyme's active site, then this was called allosteric regulation. Some allosteric effectors enhance an enzyme's catalytic activity while others diminish it.

Besides allosteric regulation of enzymatic activity, chemical modification of enzymes became an area of intense investigation. Proteins performing many different functions were found to undergo phosphorylation where one or more phosphate groups are added to or removed from proteins by enzymes called protein kinases or phosphatases, respectively. Sometimes a whole series of phosphorylation events, one after another, form a regulatory pathway or cascade.

By the early 1980s, the dogma that a protein had one biologically active shape specified by its amino acid sequence had become well entrenched. That said, it was also well accepted that the conformation of protein could change when a small molecule became bound to it. Based on this view of the protein landscape, we looked for a unique chemical modification such as a phosphate group bound to the PrP 27–30 protein that was responsible for prion infectivity. But we failed to find such a modification—had we been successful, then the production of an artificial prion might have been much easier (see chapter 17). Instead, we eventually found that the normal prion protein PrP^C refolds into a beta sheet–rich conformation to form a prion.

In refolding PrP^C into the PrP^{Sc} prion, we learned that PrP^{Sc} can fold into a multitude of different structures, many of which represent different strains of prions. Each strain or variety of prions manifests distinct biological properties such as the incubation time or the distribution of neurodegeneration (see chapters 15 and 17). We discovered that a protein can adopt multiple biologically active shapes, demonstrating the unanticipated, remarkable plasticity of proteins.

The only dark cloud that blighted our new understanding of prions was the episode with Charles Weissmann and knockout mice.

In 1990, Weissmann had suggested to me that Bruno Oesch, his former graduate student, who was now one of my postdocs, create a "knockout" mouse without the PrP gene. A knockout mouse is a genetically engineered mouse in which one or more genes have been disrupted. The object of such a study would be to find out whether a mouse that could not express PrP^C could still be infected with scrapie prions. And, of course, if this turned out to be the case, then Weissmann's cherished notion of a nucleic acid component to the prion would be strengthened. Since knocking out genes in Tg (transgenic) mice was a relatively new technique at the time, I thought such studies could wait for a year or so until the methodology was refined. Our Tg mouse studies were already giving us almost more information than we could handle. And since I thought Bruno should develop a new project unrelated to his Ph.D. research with Weissmann, I equivocated.

A few weeks later, Weissmann called to say that he would like to go ahead with knocking out the PrP gene while he was also knocking out interferon genes in mice. Since he had conducted relatively few experiments to identify a small nucleic acid in our most purified preparations but was nevertheless receiving substantial funding from our NIH Javits Center of Excellence in Neuroscience grant, I encouraged him to pursue knockout of the PrP gene.

As the methods for creating knockout mice rapidly improved, Dave Westaway periodically asked me what was happening with Weissmann's efforts. We both thought progress in Zürich was far too slow, and Dave wanted to transfer the knockout work to San Francisco, if only to demonstrate, once and for all, that no RNA was involved in our prions. I resisted that idea, because we had so much positive investigating to do, but finally in the spring of 1991 I told Weissmann of our plans to set up a parallel PrP gene knockout effort. He objected; I caved. Dave Westaway was unhappy, but I valued my relationship with Weissmann and wanted to be responsive to him. Besides, I was feeling (perhaps uncharacteristically) benign, because my exile from the biological establishment had come to an official close when the American Academy of Neurology awarded me its 1991 Potamkin Prize for Alzheimer's Disease and Related Disorders Research. The prize was bestowed in Boston, and among those present was Frank Westheimer, whose sage advice I continued to value.

That fall, the first PrP knockout mice were born in Weissmann's lab, and it seemed as though they would survive without PrP. Shortly thereafter, Weissmann and I met in Boston and laid out a set of experiments with these mice, to be done jointly in San Francisco and Zürich. Weissmann, of course, wanted to determine whether these mice could be infected with prions, but I was not nearly so enthusiastic, because I was confident that PrP was essential for prion infectivity. That said, demonstrating that PrP knockout mice could be infected with prions would force a major revision in my thinking—Chesebro and his colleagues would have been right when they had argued seven years earlier that PrP 27–30 had nothing to do with scrapie infection.

From my vantage point, I saw the PrP knockout mice as a useful adjunct for many of our studies because mouse PrP^C would no longer be a factor. I thought that PrP^C expressed in both wild-type mice and rabbits was interfering with their immune responses to injections of denatured PrP 27–30. Because the normal mice and rabbits were tolerant to PrP^C, they had difficulty recognizing even denatured PrP 27–30 as a foreign protein. Immunizing knockout mice would circumvent the tolerance problem, since the immune system of these animals had never seen PrP^C. I hoped that knockout mice would allow us to develop a large array of antibodies, especially some that would distinguish PrP^{Sc} from PrP^C. Such PrP^{Sc}-specific antibodies would be ideal for a rapid diagnostic prion test.

The next day, Weissmann introduced me to a Boston patent lawyer who had just filed a patent for him on the PrP knockout mice.[3] The inventors were Weissmann and his students. No one from UCSF was listed as an inventor, even though the studies had been supported by our NIH grant and Westaway had developed a new procedure to isolate the DNA clone used in the knockout. At the time, I did not question the validity of the patent application, because I was naïve about issues of intellectual property, and I trusted Weissmann to do what was right. Soon thereafter, the NIH itself established regulations that would have prohibited Weissmann from filing the application without concurrence from the University of California, the official recipient of the NIH grant that supported much of the knockout work.

When I told Weissmann that I assumed we would do all of the animal inoculation experiments in San Francisco, where so much effort and money had been expended for animal care and testing, he said he wanted to do some of the inoculations in Zürich so he could see the results as they emerged. Although I found his response a bit odd, I preferred not to ponder his motives. His manner seemed genuine, and I decided that any black thoughts on my part were paranoid.

We helped Weissmann establish an animal colony in Zürich so that he could perform the experiments we had planned together. Not much happened after that; at frequent intervals, Weissmann would rue the delays he was suffering due to the University of Zürich bureaucracy, and his graduate student would plead with me not to speed ahead with our somewhat parallel studies, claiming that Weissmann was keen to see the results of these "seminal" experiments for himself as they emerged. In response to all the unhappiness in Zürich, I slowed our breeding of the PrP knockout mice; we had many other demands for Tg mice in ongoing experiments.

I was well aware that there was a remote possibility that the PrP knockout mice might develop scrapie. If so, one of two explanations would be relevant: (1) the disease they developed was not really scrapie but a similar disorder caused by some transmissible agent contaminating our inocula; or (2) the disease *was* scrapie, and some other, still elusive, molecule was the infectious agent! To prevent the first possibility, I exposed our prion inocula to ultraviolet irradiation and heat to kill any viruses. I was particularly concerned about a mouse virus isolated in southern California from wild mice living around Lake Casitas. The Casitas virus causes spongiform pathology in the brain that resembles scrapie in some respects. If the second possibility proved correct, and the prion did not require PrP, then I saw the knockout mice as the perfect host from which to isolate its essential component. No longer would PrP^{Sc} obstruct our view of the prion. Perhaps this essential component was a small nucleic acid, or even another protein. If it existed, I wanted to isolate it.

In the spring of 1992, Weissmann and I published a paper describing the production of knockout mice lacking a PrP gene and noting that the mice showed no behavioral problems and had a normal lifespan. When Steve DeArmond looked at their brains, he had found no structural abnormalities.[4] Whatever problems may have been created by the loss of the PrP gene, they were subtle.

That April, I was elected to the National Academy of Sciences. An outlaw no more, I received four more awards that year, including the magnificently named Christopher Columbus Quincentennial Discovery Award in Biomedical Research from the NIH.

The experiments to test for resistance of the knockout mice to scrapie were finally under way in San Francisco and Zürich by the fall of 1992. At a symposium in Paris on CJD, I became disturbed when I heard Weissmann's redacted version of the prion story. He condensed almost a dozen lines of evidence that PrP^{Sc} is an essential component of the infectious prion into three: (1) correlation of prion infectivity and PrP^{Sc}, (2) genetic linkage of GSS and mutant PrP genes, and (3) genetic linkage of

PrP genes and mouse incubation times. His account minimized our work and inflated the importance of the knockout mouse experiments. But by late 1992, we knew from experiments in both San Francisco and Zürich that the knockout mice were resistant to scrapie. I sensed disappointment in Weissmann's voice on the telephone when he said, "Stan, everything is working out your way."

In January 1993, Weissmann, as a member of the board of directors of Hoffmann–La Roche, was invited to give a lecture on prions at a meeting celebrating the fifteenth anniversary of the founding of Genentech in South San Francisco. I picked him up at the airport, since he had recently suffered a severe skiing injury and was on crutches. When we reached his hotel room, he asked me to look at the manuscript of a paper he had written for publication in *Cell* on the resistance of the knockout mice to scrapie prion infection. With that, he hobbled on his crutches to the bathroom. I looked at the title page to find an all-Zürich byline—just like the patent application. I was stunned! Was I crazy to be upset? Then I thought, "If Weissmann thinks this is appropriate, it must be so."

When Weissmann returned, he said, "I'm troubled by this one data point, showing that prions replicated in the knockout mouse. Are your data from similar experiments ready?"

"No," I said. "We'll have the results in a couple of months. Why don't we wait for the data from San Francisco, and then we can publish together?"

"Someone else might scoop us. I want to press ahead," Weissmann said.

I was in a state of shock about the authorship, but during my drive home I worried that I was being petty or inappropriately aggressive. I was well aware that a single paper is generally not important; what usually counts in science is a steady stream of original papers. But there are notable exceptions. I talked the situation over with my longtime mentor and friend Charley Yanofsky, who advised me to break my ties with Weissmann. "He's obviously trying to grab your work," Charley said. "You never had a true collaboration, because Weissmann had little interest in the problem. Now that he's interested, he wants it all to himself."

I found it difficult to come to terms with Yanofsky's assessment, and nearly six weeks passed before I wrote Weissmann a letter expressing my concerns:

March 22, 1993

Dear Charles,

When I read your paper on the resistance of Prn-p$^{0/0}$ mice to scrapie, I felt ambivalent. On one hand, I was grateful for the glowing references to my work; but on the other, I was concerned about a number of issues. . . . Let me also say that I am extremely appreciative for all of your enthusiastic support over the last decade and for all that you have taught me. My ambivalence about the paper seems to come from my belief that these experiments with Prn-p$^{0/0}$ mice were a collaborative project. I hope that you can understand how disappointed I was when I saw that the paper had an all Zürich by-line. Although I tried to play a limited (but significant) role in these studies, I always considered them to be a collaborative effort. . . . Because I sensed your enthusiasm and pride in this work, I have tried to remain involved without being too assertive. However, I did think that I was always involved. We performed bioassays on a number of your coded samples, I provided you with whatever reagents you needed, including scrapie prion inocula, anti-PrP antiserum and breeding pairs of our Tg(SHaPrP)81 mice, and we carried out studies in parallel giving you data which I believe should have been included in the paper. Also, our joint NIH grant paid for some of these studies.

Charles, I understand your group did the bulk of the experiments and that you believe that this is an important paper for you to publish without me. I am not as concerned about this manuscript as I am about the future of our collaborative relationship. It seems to me that we now need to establish guidelines to separate collaborative projects from independent ones. I hope that we can do this because I would genuinely like to continue our productive relationship, but I am not sure where you stand. . . .

I hope you realize just how hard this letter is for me to write. I am truly grateful for all your generous help, enthusiastic support and insightful discussions over the past decade, and I genuinely hope that we can come to some arrangement that will allow us to continue our most enjoyable and stimulating collaboration. . . .

Cordially,

Stanley B. Prusiner

In response, he called to say that his laboratory was seen as an outpost of mine and he needed to reestablish his own identity.

What was I to do? Should I ignore Weissmann's behavior or cut my ties? That summer, I again discussed my dilemma with Charley, who had seen Weissmann at the Zürich Opera only a few weeks earlier and noted that the mention of my name brought an unhappy look to Weissmann's face. Charley encouraged me to push harder. "Strange things can happen in science," he said. "You cannot afford to sit back on your laurels and let someone else steal your work."

I knew Charley was right: I needed to end my professional relationship with Weissmann. I had wasted too much emotional energy on this situation already. So I wrote Weissmann another letter:

November 8, 1993

Dear Charles,

Had I been able to separate more clearly my warm feeling of friendship and admiration for you from the "business" of academic science, I would have discussed this issue with you before our January, 1993, meeting when I viewed your draft of the *Cell* paper with disbelief.

Having said all this . . . I think that it is time to put friendship aside and deal with the business of science. It seems to me that at this time there is no advantage for either of us to continue collaborating in the foreseeable future. You see your collaboration with me as a negative with respect to your own recognition and credit. I see our collaboration as a problem that I do not know how to fix.

On the other hand, I would hope that our agreement not to collaborate for the time being might preserve our friendship and allow us to collaborate in the future if a situation arises where we see a mutually beneficial arrangement. . . .

I hope that you understand my feelings and point of view. I am sorry if this proposed change in our relationship presents a problem, but I am convinced that it would be unwise to continue on the present course. . . .

With best wishes,
Stanley B. Prusiner

My elation over the results of the knockout mouse experiments was all but destroyed by my sorrow over the loss of a friend. I had enjoyed interacting with Weissmann; I appreciated his Talmudic approach to science and I was always delighted to have his penetrating critiques. But our friendship was over, and it seemed unlikely to be revived.

14

Turf Battles

It often happens, with regard to new inventions, that one part of
the general public finds them useless and another part considers
them to be impossible. When it becomes clear that the possibility
and the usefulness can no longer be denied, most agree that
the thing was fairly easy to discover and that they knew it was
significant.

—Abraham Edelcrantz, *Treatise on Telegraphs* (1796)

By now, the opposition among my colleagues to the prion had
all but collapsed, and our work was about to begin to receive
significant public recognition. There were a couple of draw-
backs to my newfound fame: The less serious was the attention
of the press, which I had managed to keep at bay for the past eight years.
The other, considerably more disturbing to me, was the steady encroach-
ment—or perhaps I should say poaching—on my territory by the formidable Carleton Gajdusek.

But first, to the press: In June 1994, I was in Paris for a monthlong
appointment at the Collège de France, during which François Jacob had
arranged for me to give a series of eight lectures on prions. While there, I
received a phone call from Jordan Gutterman, a professor at the M. D.
Anderson Cancer Hospital in Houston, who, as head of the awards

committee at the Lasker Foundation, announced that I had been given the 1994 Albert Lasker Award for Basic Medical Research.

In late September, I flew to New York a few days before the awards ceremony, which promised to be an extraordinary occasion. The award is a prestigious one and consisted of a $25,000 honorarium, a citation, and an inscribed statuette of the Winged Victory of Samothrace, symbolizing conquest over death and disease. I was especially happy to learn that John Clements, a pediatric pulmonologist at the UCSF Cardiovascular Research Institute, was due to receive the Lasker Award for clinical medical research, and that the genetics pioneer Maclyn McCarty, one of my earliest supporters, would receive a Special Achievement Award.

The Lasker Foundation called me ahead of time to say that the *New York Times* wanted to do a full-page biographical article on me, with a Q & A interview. "Absolutely not," I said. Jordan Gutterman then called to say that I had to talk to the *Times*. The Lasker award was newsworthy, he said, and it would be unthinkable not to. I explained that news stories about me and my work were uniformly negative and so odious that I preferred to avoid them. He pleaded, and I finally caved in. I stipulated that any interview would have to focus on the science, and I gave him a time when I would be available. At the appointed hour, I went to the old *New York Times* building on 43rd Street and took an elevator to one of the upper floors, where the science writer Gina Kolata met me. A slender woman with dark hair and an apparently warm personality, she led me past row after row of desks where reporters sat, into the office of her editor, who was away. The first question out of her mouth was about my personal background. I undoubtedly projected annoyance since I had been so direct earlier about not wanting to discuss the story of my life. "I'm here to speak about the science, not myself," I snapped.

Following a few perfunctory questions about prions, she shifted gears: "How do you answer your critics?" she wanted to know. At that point, I realized her article would be inflammatory—and that no matter what I said, she was going to focus on the objections to our work. I told her that the published record spoke for itself and that she should read my papers. This was, I knew, no way to speak to a journalist, so I tried to summarize the data for her, but she didn't seem interested. The interview was over in twenty minutes and I was happy to leave. It had turned out as badly as I thought it would.

That evening, a dinner was held in a private room on the second floor of La Grenouille for the Lasker winners, the officers of the Foundation, and members of the selection jury. At a reception before the dinner, one of the jurors, Joseph Goldstein, a molecular geneticist at the University of Texas Southwestern Medical Center, walked over to Mac McCarty and me to congratulate us. He and Michael Brown had won the Lasker Award a decade earlier, followed by the Nobel Prize in 1985 for their pioneering work on cholesterol metabolism that led to the statin drugs; Mike Brown had been two years ahead of me at Penn and we had worked together in Earl Stadtman's laboratory. Goldstein wanted to discuss a curious parallel in careers: "Mac, you spent considerable effort showing that your preparations were composed of DNA and devoid of protein," he said, "and Stan, you've spent your career showing that your preparations are composed of protein and devoid of DNA."

"We've both been aware of this remarkable symmetry for a while," I said, "and we've been thinking of writing an article about it." Indeed, Mac and I worked on such a paper from time to time over the next decade. We spent many hours going over the seminal experiments he did in Oswald Avery's laboratory. We enjoyed being together, so assembling the manuscript never felt like a chore. There was a special warmth that radiated from Mac, who died in January 2005 at the age of ninety-four. I always felt privileged to have spent time with him. The paper was published in *Annual Review of Genetics* the following year.[1]

The Lasker Award ceremony was held the next morning at the Pierre Hotel, and a brief note describing all three winners was published in the *New York Times*. There was no article by Gina Kolata; maybe I had dodged another bullet. Could it be that she had read a few of my papers and realized that the science was solid? Presumably, she was smart enough—after all, she worked for the *New York Times,* and they were thought to have the best and the brightest.

At 7:00 A.M. the following Tuesday, when I was back home in San Francisco, the telephone rang. It was Jordan Gutterman again. "Have you seen the Science section of the *New York Times*?" he said.

"No. Why do you ask?" I replied.

"It's terrible! You warned me, but I wouldn't listen. To quote Jean Monnet, 'You may think you understand something in your head, but you really won't know it until it's in your bones.' Well, I now feel it in my bones."

His bones? I told him I was sorry he was upset. I could tell from his tone that he was worried that the Lasker Foundation had made a dreadful mistake. Did he want his award back? I had been given it publicly, so if he did there would be a rumpus. He said he would call me later. I hurriedly dressed and went out to buy a copy of the *Times*.

The headline, across the bottom third of the front page of the *Times*'s Science section, read: "Viruses or Prions: An Old Medical Debate Still Rages." To the left was a picture of me stepping down from the dais holding my Winged Victory; to the right was a subhead reading "Lasker Jury Takes One Side of Controversy." My interpretation of the article by Gina Kolata was that it implied that my Lasker award was something of a surprise to the science community, noting:

> For more than a decade, Dr. Stanley B. Prusiner, a neurologist at the University of California at San Francisco, and his collaborators have zealously promoted [an] iconoclastic theory in the face of skepticism and hostility from many scientists. Dr. Prusiner says that there is a class of unusual proteins, which he calls prions, that can cause certain neurodegenerative diseases by inducing normal proteins to change their configuration and become deadly.[2]

Gutterman, as head of the awards committee, was quoted as saying, "Obviously, the Lasker jury did not feel that this is controversial. Even if the final proof is debatable, some of the most outstanding minds today think this is as solid as it can get." Next came a line of praise from Phillip Sharp, a Lasker and Nobel laureate and chair of MIT's Biology Department ("one of the most remarkable developments in infectious diseases that I know of"). I was described as "a genial man with a shock of curly gray hair and a ready laugh" and given a few paragraphs to state my case. After that, fully half the article was given over to several of my critics, including Heino Diringer and David Bolton, who explained why my work on prions must be wrong. The article concluded with some musings from Gajdusek's erstwhile postdoc Robert Rohwer, now director of the molecular neurovirology unit at the Veterans Affairs Medical Center in Baltimore, who was earlier quoted as saying that he had "spent a good part of my career looking at some of these outrageous claims and they don't hold up."

"My view of this whole field is that in the end it will be of greatest interest to sociologists or psychologists of science," he said. "There's a sort of boredom with virology. People want to be entertained by science." And the prion hypothesis, he said, "is very entertaining."

A few hours later, Gutterman called me back to say that he had just spoken with Joseph Goldstein and David Baltimore, and they had assured him that the Lasker Foundation had not made a mistake. Rather, he said, the award to me should be seen as "reflecting the courage of the Foundation."

How much of this story had to do with my impatience with the press and, more particularly, Gina Kolata's questions? The answer is probably unknowable, but I see the experience as just another example of a reporter intent on "capturing the controversy." She was writing a great story—one that would help sell papers, one that her editors liked. She and her editors seemed to believe that she had uncovered the truth about prions, while the Lasker jury, composed of eminent scientists, had been fooled.

As it happened, there was more to the story than what Kolata reported. A choir of imitators had slowly been drowning out the refrain of the naysayers. While I was delighted that the prion concept was beginning to be embraced, I was nettled to learn that others were claiming credit for several ideas that sprang from my work. Suddenly, the very virologists who had vociferously attacked my findings were sidling up close. From the Wistar Institute in Philadelphia, Hilary Koprowski sent me a slide entitled "Four Stages of Adopting a New Idea."

- The first is, "It's impossible, it's nonsense, don't waste my time."
- The second is, "Maybe it's possible, but it's weak and uninteresting. It's clearly not important."
- The third is, "It's true and I told you so. I always said it was a good idea."
- And the fourth is, "I thought of it first."

Although I thought the published record on the discovery of PrP 27–30 was unassailable, I was mistaken. Back in 1985, I had been startled to read, in a chapter by Carleton Gajdusek in a new virology textbook, that he claimed to have developed a thousandfold enrichment of the scrapie agent and that proteases diminished its infectivity.[3] He wrote,

"Rohwer and Gajdusek succeeded in a 1,000-fold purification of scrapie virus relative to other quantifiable proteins in the original brain suspension. In such preparations the virus is susceptible to proteinase K and trypsin digestion, but it is not inactivated by nuclease." Accompanying this assertion was a reference to a 1980 paper they had published in the proceedings of a symposium on multiple sclerosis held in Frankfurt, Germany.[4] These were virtually similar to my results—the very results that had led me to introduce the prion hypothesis in *Science* in 1982. Because the 1980 Rohwer and Gajdusek reference predated my work by a couple of years, I was horrified. As it turned out, I had sent one of my colleagues to that Frankfurt meeting, so I had a copy of the symposium proceedings—and, of course, there was no mention of the work that Gajdusek had attributed to that publication.

I was unsure how to mount a counterattack. When I visited Erling Norrby in Sweden a few weeks later, I showed him Gajdusek's chapter. He suggested that I write Gajdusek a letter describing my distress and demanding a public retraction. I sent the letter by certified mail, but it elicited no response. After a few more weeks, I called Gajdusek to voice my concerns. He responded by claiming that he had discovered PrP.

"That's simply not true," I protested. "The path of discovery is clearly documented in papers published in well-respected, refereed journals."

"I discovered the protein, and it is described in my notebooks long before your papers were published," he said.

"Unpublished data in notebooks do not constitute a legitimate scientific discovery," I pointed out—but Gajdusek wanted no part of my argument. In his mind, a 1959 paper of which he was a coauthor, describing amyloid plaques in the brains of patients with kuru, was all he needed to justify his claims.[5] This scenario was all the more remarkable when I considered how long he had insisted that a virus caused scrapie, kuru, and CJD.

"I'm going to hang up now," I said. "This conversation is not productive, and I'm wasting money." After more than forty minutes on the telephone, I could not bear to listen to him any longer.

My only recourse was to write the editor of the textbook, asking for some kind of retraction. The editor, Bernard N. Fields, responded that he was sorry about the problem but unwilling to rectify the situation. He seemed uninterested in my plight; moreover, he probably thought that my protest had little merit since he, too, was convinced that a virus caused scrapie. (Eventually he would change his mind, and we became friends two years before he died of pancreatic cancer in 1995.) I contemplated lodging a formal complaint with James Wyngaarden, the NIH director, but I decided against it. Gajdusek had a Nobel Prize, and I feared that this accolade would give him undue power in any struggle I might mount.

Gajdusek continued to give lectures and write papers throughout the late 1980s and early 1990s calling PrP 27–30 the "scrapie amyloid protein," which he claimed to have discovered. This would frequently generate letters and telephone calls from friends telling me what had transpired. They invariably wanted me to stop him, but I would reply, "I think you can do more to stop him than I can. I've confronted him several times, but I'm unable to alter his behavior." On more than one occasion, Lee Hood sensed my distress. "Everyone knows where the work on prions was done," he said one day in 1991 as we were driving across Pasadena to his Caltech lab. "No knowledgeable person thinks Gajdusek did the work. The only person who thinks Gajdusek made these discoveries is Gajdusek."

One of the more unsettling moments in this battle came in early 1996, when Erik Lycke telephoned me from Göteborg. "The people in Stockholm"—shorthand for the Nobel Committee—"are unclear about Gajdusek's role in the discovery of prions," he said. "Gajdusek is telling them that he discovered PrP, which he calls the 'scrapie amyloid protein.'"

This was a nightmare. I could see the headlines: GAJDUSEK WINS SECOND NOBEL FOR HIS DISCOVERY OF PRIONS. "Erik, the path of discovery is well documented in peer-reviewed journals."

"I know that, I've read them. Please do me a favor. Write me a letter documenting the discovery with references." While he did not say the letter was for the Nobel Committee, it was easy to surmise that this was the case. Erik had already nominated me for the Nobel Prize several times, so I assumed that he was involved in some ongoing analysis. Lycke and I first became friends when we visited Israel in 1978. Erik was a superb scientist whose remarkably broad interests stretched from virology to neurochemistry. He was always warm and kind to me as well as being incredibly supportive. At his retirement gathering in Göteborg in 1992, he embarrassed me when he said, "Unlike most of us, Stan has really accomplished something. He has discovered prions, which are an entirely new class of infectious particles."

My worries over Gajdusek's attempts to take credit for my work ended shortly thereafter when the FBI arrested him in April 1996 for child molestation. He was formally charged with having molested one of the fifty-seven children from New Guinea and Micronesia he had adopted and brought to the United States to be educated. After a plea bargain in February 1997, he was sentenced to between nine and twelve months in jail and five years of unsupervised parole—provided he spent that time abroad. When he was released from jail, his friends threw a party for him at Dulles International Airport, where he boarded an airplane for Paris. Six months later, I caught up with him at a meeting held in a resort at the Dead Sea in Israel. The biophysicist Ephraim Katzir-Katchalski[6] and Albert Taraboulos, one of my former postdocs and now on the faculty of the Hebrew University in Jerusalem, had organized the meeting. During a coffee break, Ephraim asked me how I came up with the idea of prions.

As I told Ephraim about the steady accumulation of data arguing for infectious proteins, Gajdusek, seriously overweight, clad in a checked shirt and overalls, and appearing (understandably) somewhat depressed, joined our conversation. I acknowledged him by noting that many people, himself included, had contributed data that led to the idea of scrapie being caused by a protein.

"I told you that the scrapie agent was a protein when you visited me in New Guinea in 1978," Gajdusek bellowed.

"I'm sorry, but I don't recall you telling me that it was a protein," I said.

"It's all in my diaries!"

"I don't remember any such conversation," I reiterated.

"If you thought it was a protein," Ephraim asked him, "then why did you call it a virus?"

"I *still* call it a virus," Gajdusek shouted, defiantly shaking his finger in Ephraim's face.

"I don't think it's productive for me to say anything about these issues, so I'll leave the two of you to converse," I said, and backed off as politely as I could.

I saw Gajdusek on many occasions over the next decade at various meetings in Europe, where he lived and traveled widely during his self-imposed exile after his release from prison. The last time was at a meeting at the Royal Society in London in September 2007, commemorating the end of kuru. The last case had occurred in 2004. Gajdusek appeared delighted to see me and seemed eager to tell me that he had recently emerged from a coma that had lasted forty-eight days. (Michael Alpers later told me that Gajdusek's coma was forty-eight hours in length and had been caused by congestive heart failure.) Gajdusek's massive ankle edema and poor cardiac status severely limited his mobility. Still overweight and wearing one of his checked shirts, he was unable to stand long enough to give a short talk. Evidently unrepentant about the offenses of which he had been convicted, he spoke at length about the penises of young boys, to the disgust and embarrassment of many in the audience. About a dozen Fore attended the meeting, and they must have brought at least some measure of joy to Gajdusek, since their view of "Doctor America" remained unchanged.

Gajdusek spent much of his time living in a hotel in Tromsø, Norway, near the Arctic Circle. One morning in December 2008, at age eighty-five, he was found dead in his hotel room. Sadly, he died alone—none of his fifty-seven adopted children was present during the final hours of what had been an extraordinary life by any measure. I was told that his ashes would be buried next to his sister in Bratislava, but later learned that they were scattered in the Arctic Ocean.

15

Mad Cow and Other Prion Strains

The need for a better understanding of the prion's molecular structure was dramatically apparent during the third week of March 1996. The British scientist Robert G. Will and several of his colleagues in the United Kingdom's National Creutzfeldt-Jakob Disease Surveillance Unit (NCJDSU) were suddenly summoned back to London from a prion meeting in Paris. They had written a paper describing ten young patients with CJD that was about to be published in the *Lancet,* and the government felt compelled to acknowledge the possibility that these patients had contracted prion disease from eating tainted British beef.[1]

The NCJDSU had been founded six years earlier to monitor any changes in the incidence of Creutzfeldt-Jakob disease in the wake of Britain's BSE (bovine spongiform encephalopathy) epidemic. Three of the ten CJD patients whose cases were chronicled by Will and his colleagues were teenagers, and the others were all under forty. CJD usually appears in much older people, the rare exceptions being the young who contracted the disease from contaminated human growth hormone taken from cadavers of victims of the disease. The NCJDSU scientists had performed neuropathologic studies on these ten patients and found large amounts of PrP amyloid surrounded by a halo of spongiform degeneration. Notably, the neuropathologic changes in young people who received tainted growth hormone differed from these lesions. This form of CJD seemed to be a new disease. In their *Lancet* paper, the authors labeled it "new variant CJD," or vCJD.

The British government's fears that this disturbing news would leak were realized on the morning of March 20, when the London *Daily Mirror* carried a front-page article headlined "OFFICIAL: MAD COW CAN KILL YOU" and announced that "Humans could catch Mad Cow Disease from eating infected beef, the government will admit today." On page 2, the headline read "WE'VE ALREADY EATEN 1,000,000 MAD COWS," and was followed by a longer text:

The government launched a "don't panic" campaign last night....

Health Secretary Stephen Dorrell's announcement today that people could catch the human version from eating infected beef will have a devastating effect on the food industry.

A national advertising campaign will start tomorrow aimed at reassuring people that eating beef is not dangerous.

But some experts fear we may already have consumed more than a MILLION infected animals—and that as the incurable killer disease can take 10 years or more to show up, some people are already living on borrowed time.

Ministers have insisted for 10 years that it was impossible for it to jump from cows to humans.

But Mr Dorrell plans to tell MPs that after an upsurge in human cases, Whitehall scientists now fear it might be transmitted from species to species.

He will maintain there is no clear link and that deaths here from the human form are at similar levels to other European countries. But doctors diagnosed 55 new cases in 1994—13 more than in the previous year and double the rate a decade ago.

Sales of beef pies, sausages, mince, joints and other products are expected to plummet, in the wake of Mr Dorrell's announcement.

Schools in many areas... already ban beef from their menus and others are likely to follow.

And other countries could ban British beef to stop the disease spreading.

Getting It Right

Bovine spongiform encephalopathy, or mad cow disease, had first been reported in Britain ten years earlier and linked to the preparation of feed supplements rendered from the carcasses of livestock. The disease was subsequently reported in ten other countries and linked to cattle feed from the United Kingdom, and the use of offal in cattle feed was banned there on June 14, 1988. A link between BSE and certain cases of CJD had long been suspected, but the first open acknowledgment by the British government that the connection was "likely," and that these infections had preceded the ban on offal, came only on the heels of the leaked news report in the *Daily Mirror*.

Following the government's announcement, the European Union imposed a worldwide ban on all exports of British cattle and beef products. The international press went into high gear, focusing on mad cows and the young people they had killed. The headline above a picture of a British cow on the cover of the April 1, 1996, issue of *Time* magazine was "CAN THIS COW KILL YOU?" Two weeks later, Oprah Winfrey opened her daily afternoon TV show with an apocalyptic announcement:

> Mad cow disease. It's a medical mystery spreading panic across the Atlantic. In England, ten puzzling deaths of young people may be linked to a rare and fatal brain disorder in cattle. The afflicted cattle shake and contort like mad dogs before what must be an excruciating and inevitable death. In human beings, dementia and paralysis precede death. Scientists speculate that cattle contract the disease by feeding on sheep parts that are infected with another disease—a practice officially banned in England. . . . Could it happen here?

And she continued: "It has just stopped me cold from eating another burger!"

Oprah's decision not to eat another burger prompted the American Cattlemen's Association to bring a $10.3 million lawsuit against her, claiming that her remarks had forced already slumping beef prices to ten-year lows within a week of the show. The jury eventually ruled in favor of Oprah. "Free speech not only lives, it rocks," Oprah told reporters. "I'm still off hamburgers," she added.

The first question, of course, was whether the new variant disease had really come from prion-tainted beef. If so, how many more such cases were out there? And where else could we expect variant CJD to appear, besides Britain? The answers to these questions were frustratingly vague, but alarm grew after the *Daily Mirror*'s article on March 20, 1996, in which investigators at Oxford and the Central Veterinary Laboratory in Weybridge, Surrey, were quoted as saying that almost a million British cattle may have been infected with BSE prions, and most of these had already entered the human food chain. This was six times more than the 161,000 cattle that had been diagnosed with BSE over the previous decade. After completing their speculative analyses, Roy Anderson and his colleagues published their million BSE cattle estimate in *Nature*.[2]

Gerald Wells and his colleagues at the Central Veterinary Laboratory were the first to describe spongiform degeneration in the brainstems of dying British cattle, in a 1987 paper.[3] Because the microscopic changes they found were similar to those seen in scrapie, some veterinarians at first called the disease "bovine scrapie," but in their paper Wells et al. labeled it "bovine spongiform encephalopathy." It was the press, always on the lookout for a memorable catchphrase, that named the disorder "mad cow disease." The neuropathologic changes Wells found in the brains of the cattle suggested that BSE was a prion disease, but another year would pass before a study was published showing that PrP^{Sc} was indeed present in the brainstems of cattle with BSE.[4]

In May 1988, the Ministry of Agriculture, Fisheries and Food under John MacGregor, appointed a committee to study BSE and make recommendations to the British government about how to contain it (figure 29). The committee was known as the Southwood Working Party, for its chairman, Oxford zoologist Richard Southwood.[5] With the help of considerable data from Wilesmith on the location of each British cow and bull with BSE, the committee concluded that the cattle were becoming infected with scrapie prions in their feed supplements—meat-and-bone meal that had been rendered from dead livestock—and that these supplements should no longer be fed to cattle. Only about half of a typical slaughtered farm animal went for human consumption; the remainder, called offal, was processed in rendering plants into high-protein and high-fat fractions, with the latter turned into lard, cosmetics, soap, and (most alarming) baby food. Most of the protein fraction became meat-and-bone meal food pellets, which were fed to farm animals. On February 3, 1989, the committee made its recommendations. ("And we got it right," John Walton later remarked to me.) These were: that cattle with clinical signs of BSE be destroyed and kept out of the human food supply; that meat-and-bone meal and other dietary supplements from ruminant

Figure 29   The U.K. minister of agriculture John MacGregor is generally credited with controlling the spread of mad cow prions by a series of visionary feed bans.

animals no longer be fed to cattle; that milk from cattle with clinical signs be destroyed; that ruminant offal, including brain, spinal cord, and lymphoid tissue, should be excluded from baby food; and that bovine products should be excluded from medicines such as vaccines. Their only misjudgment was in noting that "the risk of transmission of BSE to humans appears remote."

Even before these recommendations were issued officially, Agriculture Minister MacGregor had held discussions with the committee and instituted an immediate policy. In August 1988, he ordered the slaughter of all cattle showing signs of the nervous system disease. MacGregor was caught between his concern for food safety and the need to protect British farmers from economic ruin: Was he overreacting and thereby costing the United Kingdom millions of pounds sterling? If he did nothing, was he putting the safety of the British food supply at risk?

The geographical distribution of BSE was puzzling. According to John Wilesmith's records, the disease was much more prevalent in southern England than in the northern United Kingdom. Had a new virulent strain of the scrapie prion arisen in sheep in southern England and contaminated the meat-and-bone meal? Even more puzzling, since scrapie had been endemic in British sheep for hundreds of years, why, just now, were cattle developing a scrapielike disease? Meanwhile, the number of infected cattle was increasing each month. On July 18, 1989, the European Union banned the export of British cattle born before July 1988 and the offspring of infected or suspect animals. Mad cow disease was a constant topic in the news and rapidly becoming an economic and political disaster. The losses suffered by British farmers, abattoirs, renderers, and butchers were staggering. Cattle in Scotland were deemed untainted, and Scottish and imported beef was sold at a premium in butcher shops and restaurants elsewhere in Great Britain. On the menus of airlines flying in and out of Britain, the non-British origin of the beef was clearly indicated. In the back of everyone's mind was the worry that BSE prions might be transmitted to people and eventually cause CJD.

On further advice from the Southwood committee and agriculture officials, MacGregor issued a Specified Bovine Offal ban in June 1989, which prevented brain, spinal cord, and lymphoid tissues from cattle from entering the human food supply, an extension of his earlier directive banning it from being used in baby food. In a routine midterm cabinet shuffle later that summer, Prime Minister Margaret Thatcher named MacGregor minister of education and science and appointed his deputy, John Gummer, as minister of agriculture. One of Gummer's jobs was to assure the public that British beef was safe to eat. At a county fair the following May, the press cornered him and goaded him into being photographed with his four-year-old daughter, Cordelia, both chomping on hamburgers, which more than a few people ridiculed as "forced feeding."

In 1991, Wilesmith's epidemiological sleuthing finally uncovered the explanation for the outbreak of BSE—industrial cannibalism.[6] In the 1970s, a number of events coalesced to alter the rendering process in many parts of England. The Arab oil embargo of 1973–74 had drastically increased the cost of energy needed to heat the rendering vats. Repairs and increased insurance rates after a 1974 explosion at a chemical factory at Flixborough had resulted in higher prices for organic solvents. At the same time, people were turning away from the use of lard in cooking and beginning to substitute vegetable oils. With a surplus of tallow developing, the price fell. As all these economic forces were demanding changes in the rendering industry, the British government let regulations on that industry expire, as part of its program to deregulate the economy. To accommodate the economic changes, the renderers decided to reduce the temperature of the rendering process and omit the organic solvents. As a result, when the offal was cooked, not all of the fat floated to the tops of the vats, so the protein fraction contained some fat suspended in the liquid. From this, high-fat meat-and-bone meal pellets were milled; the new meat-and-bone meal contained about 14 percent fat, whereas the old meat-and-bone meal had contained only about 1 percent fat and had required the addition of exogenous fat when it was formulated into pellets so that livestock would find it palatable. The renderers were now producing an edible meat-and-bone meal using less heat and no organic solvent; its reduced cost was a clever solution to the economic changes they were encountering. No one foresaw the coming disaster.

The proportion of meat-and-bone meal processed with organic solvents fell from at least 70 percent in the mid-1970s to about 10 percent in the early 1980s. This change in solvent use is thought to have occurred at the same time as the first BSE prion transmissions to cattle. The north-south gradient of BSE incidence appears to correlate with the use of solvents: In Scotland, organic solvents continued to be used, and the incidence was seven times lower than in southern England, where they were not.

The origin of the BSE prions themselves remains unclear. One hypothesis holds that they arose spontaneously in cattle, like sporadic CJD in people, and another argues that they came from sheep. Rare cases of BSE in North America and Japan over the past decade—particularly those in young cattle—support the first hypothesis. The endemic nature of scrapie in British sheep for almost three centuries argues for sheep prions being the culprit. Whatever the source, almost ten years passed before the prions that survived the new rendering process began to produce recognizable disease in cattle. As the initial small number of bovine or sheep prions stimulated conversion of bovine PrP^C into bovine PrP^{Sc}, bovine prions started to multiply in large numbers. Since most of the cattle were slaughtered by the time they were two years old, they never manifested signs of neurological dysfunction—but many of them may well have had bovine prions in their brains and possibly in other tissues when they were slaughtered. Only dairy cattle, which typically live beyond three years of age, manifested neurologic disease, at first appearing apprehensive and later exhibiting difficulty walking. In addition, they often manifested an exaggerated startle response to loud sounds or other stimuli. (The exaggerated startle response in BSE cattle resembles startle myoclonus in CJD patients, which is also frequently elicited by a loud noise.) Just as ritualistic cannibalism caused kuru, industrial cannibalism created mad cow disease. The repeated rendering of prion-infected offals appears to have selected for a strain of bovine prions that is pathogenic for humans.

Since the human disease was first reported in 1996, more than two hundred teenagers and young adults have developed variant CJD. Most of the cases have been in Britain, but small numbers have also been seen in ten other countries, including the United States. Since 2005, however, fewer than six variant CJD cases annually have been recorded in Britain. It is likely that the measures instituted by John MacGregor and his colleagues at the beginning of the BSE crisis saved thousands of lives. While there is little to cheer about with regard to BSE, the feed and offal bans of the late 1980s seem to have been key in containing the spread of prions both to cattle and to people.

Looking for Retribution

At the same time that bovine prions were causing an epidemic in cattle as well as infecting people, human prions in tainted human growth hormone were beginning to manifest CJD in young people who had been treated for small stature many years earlier. The story begins in 1921, when Herbert Evans and Joseph Long, of the University of California, Berkeley, reported growth-stimulating activity in extracts prepared from the anterior lobe of the rat pituitary. Two decades later, after suitable protein-purification techniques were developed, Evans and Choh Hao Li were able to purify growth hormone from bovine pituitaries. But attempts to use it to alleviate dwarfism in children failed, because the amino acid sequences of growth hormone from different species are not identical. In the late 1950s, Li's laboratory and two others reported the isolation of growth hormone from pituitaries taken from human cadavers, and this hormone was eventually used to treat children suffering from dwarfism and other forms of pituitary malfunction, including the simple failure to grow to normal height.

As use of the hormone escalated in the 1960s and 1970s, the demand increased, requiring the collection of more and more pituitaries from cadavers. In the United States alone, some ten thousand children were treated with cadaveric growth hormone between 1963 and 1985, when a biosynthetic human growth hormone was developed and the practice of collecting the hormone from cadavers was banned. These patients developed into adults of normal height; however, a small fraction developed CJD. More than two hundred cases of CJD in young people who received cadaveric growth hormone have been recorded worldwide.

When the initial reports on the transmission of CJD from humans to apes and monkeys were published in 1968, endocrinologists did not sound the alarm. Presumably, the disease's infrequent occurrence—recall that sporadic CJD occurs in about one person per million—and the vague nature of its infectious agent, set against a background of increasing demand for human growth hormone, conspired to nudge investigators and physicians to ignore a potential problem. Besides, there was no alternative treatment for hypopituitarism at that time.

Although several scientists in the late 1970s and early 1980s, including Alan Dickinson and Carleton Gajdusek, raised the possibility that proteins extracted from the human brain might be contaminated with the CJD agent, no experiments were performed to test such a hypothesis. Even with the introduction of the prion concept in 1982, the idea of infectious proteins remained controversial for another decade, prompting people in allied fields to plead ignorance when trying to evaluate potential problems. It was not uncommon to hear physicians and scientists alike say, "If the people in the prion field can't agree on the nature of the infectious agent, how can we be expected to make informed decisions?"

On May 30, 2002, the French minister of health appointed an international committee to investigate the high number of CJD cases in young people who had been treated with human growth hormone derived from human pituitary glands. My friend Henry Baron chaired the committee, of which I was a member. That fall, the committee met twice in Paris to discuss why the frequency of CJD in human growth hormone recipients in France was so much higher than in the United Kingdom and the United States. The committee concluded that the principal cause of CJD in human growth hormone recipients resided in pituitary gland collection from people hospitalized on neurological, neurosurgical, or psychiatric services.

This older, neurologically impaired population was considered to have a much higher likelihood of CJD than that of the general population. In other words, this group of patients might have developed CJD prions at a much higher rate than individuals in the general population. The committee was unable to identify any fault in the purification practices of Dr. Fernand Dray, who had been in charge of purifying the human growth hormone from large pools of pituitaries in a laboratory at the Pasteur Institute. None of the purification practices would be expected to concentrate the prions preferentially.

In March 2008, Henry Baron and Alice Dautry, the president of the Pasteur Institute, asked me to come to the Palais de Justice in Paris to testify on behalf of Fernand Dray, who, at age eighty-five, was being prosecuted for his extraction of growth hormone from the pituitary glands of cadavers in the 1980s. Besides Dray, there were six other defendants: three from the Association France-Hypophyse, who collected the pituitaries from cadavers, and three from the Central Pharmacy, who dissolved the freeze-dried powder containing the hormone and dispensed it into vials for distribution to pediatricians. Almost 1,700 French children and young adults were deemed at risk of contracting CJD from the treatment, and 125 had so far died of the disease.

My sympathies were with Dray and his codefendants. No one could have anticipated this scenario. On the other hand, the parents were justifiably distraught; they had agreed to, and sometimes requested, growth hormone treatment for their children. Undoubtedly, many of them now felt guilty about their decision; shifting part of the blame might bring some small measure of comfort.

During my testimony, I was asked several times about whether or not my discovery of prions had had any impact on thinking about the safety of growth hormone: "Wasn't it well established, before your work, that CJD is a transmissible disease?"

"Yes, but the nature of the transmissible CJD agent was unknown," I replied. "Information about the CJD agent was limited, because measurements had to be performed in apes and monkeys. The incubation times in these primates was one and a half to two years, and thus little information was available about the physical properties of the CJD agent."

"But it was still transmissible?"

"You are correct," I replied, "but I think you fail to understand how primitive our knowledge was at the time. Although Gibbs, Gajdusek, and Latarjet published a paper in 1978 on the extreme resistance of the CJD and kuru agents to inactivation by X-rays, the interpretation of their data was unclear.[7] For another decade, both Gibbs and Gajdusek continued to speak and write about the CJD virus. Moreover, it was not until 1985 that we published a paper demonstrating that prions cause CJD."[8]

After my response was translated into French, I continued, "It is important to realize that purified protein, in the form of insulin, had already been injected into thousands of people for several decades before endocrinologists began treating dwarfs with growth hormone. Properly prepared insulin is free of bacteria, viruses, and fungi, such as yeast; and thus it does not produce abscesses at the sites of injection or systemic infectious diseases. And the same is true for properly prepared growth hormone. To my knowledge, the growth hormone prepared by Fernand Dray was similar to other protein hormone preparations, in that they produced neither abscesses nor other acute illnesses."

Near the end of my testimony, I told the court that I had never imagined that some preparations of growth hormone might be tainted with prions. Despite being a neurologist and working on scrapie every day from 1974 on, I could not recall ever thinking about the possibility that growth hormone given to children might cause CJD. For this lack of vision, I apologized to the parents of all the children who died of CJD after receiving growth hormone. Scientists are poor at predicting the future, and this humbling experience was no exception.

I remain hopeful that my own inability to foresee prion contamination in preparations of cadaveric growth hormone will bring about a more realistic view of what is reasonable to expect from scientists. I spent the decade preceding the first report of growth hormone–caused CJD seeing patients with neurological diseases and purifying the scrapie agent.

The tragedy of growth hormone–induced CJD was somehow magnified in the Parisian courtroom. The ghosts of the 125 young French victims seemed to haunt this majestic hall. The anguish of their parents, who consented to and even encouraged the daily administration of the hormone, was palpable. Though I comprehended the magnitude of the tragedy, I could not understand how any judge or jury could demand that Dray and his six codefendants be clairvoyant. How could the magistrates require anyone to see disaster on the horizon, when there was no precedent? The trial was a Kafkaesque nightmare. Fortunately, on January 14, 2009, the tribunal found all seven defendants innocent of all charges.

Fifteen months after Dray and his codefendants were declared innocent, Henry Baron and Alice Dautry asked that I return to Paris to testify again on Fernand Dray's behalf. The families of the young French CJD victims still hungered for retribution, and they had left no stone unturned—they managed to bring new charges against Dray and Elisabeth Mugnier, a pediatrician in charge of collecting the pituitary glands and monitoring treatment. Dray and Mugnier were now being tried for involuntary manslaughter—if convicted, they would face jail sentences of up to three years. A third codefendant died by the time this new trial was under way. In October 2010, I was called again to the Palais de Justice in Paris, where I testified for three hours without a break. Just as I had testified two years earlier, I said that neither Dray nor his colleagues could have anticipated the infection of growth-hormone recipients with CJD prions. I certainly had never entertained the possibility that human-growth-hormone preparations derived from human pituitaries harbored prions before publication of a 1985 report of CJD in a twenty-year-old man from San Jose, California, who had received human growth hormone for thirteen years.[9] Bruce Berg and Tom Koch, both UCSF child neurologists, made the diagnosis of CJD in this young man, and Steve DeArmond performed the autopsy. Histologic sections revealed widespread vacuolation throughout the cerebral cortex, basal ganglia, and cerebellum.

On May 5, 2011, the court again exonerated Dray and Mugnier of all charges and preempted any future suits by the families for civil damages. The verdict stated that Dray "cannot be blamed for not having any intuition of a risk of contamination that no professional could have detected at the time." Now age eighty-nine and too sick and frail to come to court, Dray could for the first time in twenty years stop worrying about being prosecuted for a crime that he had not committed. The legal attacks perpetrated on this innocent man by remorseful parents were a tragic injustice.

Variations in Shape

Despite near universal acceptance of the prion concept by the mid-1990s, there was still an important segment of prion biology that remained to be explained. Variation among prions had long been called "strains," a term reserved for distinct populations of a particular organism—viruses, bacteria, or mice. In every case, it is changes in the genome that distinguish one strain from another. Convinced that prions were devoid of genomes (that is, of nucleic acid), I struggled for several years to develop a hypothesis that might explain the existence of "strains" of prions.

Putative prion strains were initially described by Iain Pattison, who found that goats inoculated with two different scrapie strains showed different symptoms: One strain caused an intense scratching of the neck, followed by progressive nervous system dysfunction with a wide variety of neurological signs. The other did not produce scratching but did create progressive nervous system dysfunction, with some of these goats becoming hyperexcitable and others dull and docile.[10]

I believed that the chemical structures of PrP^C and PrP^{Sc} were identical. We had expended much effort in a search for a chemical difference and had found none. So how to explain diversity among prions? Often after a lecture, I was asked that question. My invariable response was, "I don't know how to explain it. I realize that's not a satisfying answer. We need to understand how the specific biological properties of a prion strain are enciphered in something other than RNA or DNA."

When I was preparing a review on the advances in prion biology for an article in *Science* in 1991, I knew I needed to confront the strain issue. For many years, I had deliberately ignored it—my focus was on determining the structure of a single strain before trying to decipher the attributes that distinguished one strain from another—but now I felt I had to set forth at least one proposal. Since it had become increasingly likely that prions were composed solely of PrP^{Sc} molecules, I suggested that each strain of prion might represent a different shape of PrP^{Sc}.[11] But this was pure conjecture, and I did not know how to gather data that would either support the idea or overturn it.

The first inroad into unraveling the strain mystery came from Dick Marsh (see figure 8). He isolated two prion strains from mink by repeated transmission of the prions through hamsters. One strain he called "Hyper," reflecting heightened sensitivity to touch, pronounced incoordination, and truncal ataxia; the other he called "Drowsy," emphasizing the hamsters' lethargic behavior. I first learned about the hyper and drowsy prion strains when Marsh presented his studies at a memorable BSE meeting in Brussels in November 1990.

Marsh spoke about an outbreak of transmissible mink encephalopathy at a mink ranch in Stetsonville, Wisconsin.[12] The episode had been puzzling, because earlier outbreaks of TME had all been traced to the feeding of sheep carcasses, presumably infected with scrapie prions, to mink. But, he stressed, "there were no sheep within a hundred miles of the mink ranch, so what could be the source of the TME agent?"

I expected Marsh to tell us that the ubiquitous "scrapie virus" was the cause of the infection. Instead, he explained that the mink farmer acquired the carcasses of cows from a nearby dairy farm and fed them to his mink. "The TME outbreak," Marsh said, "was probably due to a sporadic case of BSE, which starts with the spontaneous formation of PrP^{Sc}. This is what is thought to happen in sporadic CJD." I was happy to hear that Dick had become a convert; he had made the journey from viruses to viroids to prions. He continued, "The scrapie-virus hypothesis is incompatible with my findings, which were readily explained by the prion concept."

I couldn't wait to ask Dick what made him change his views. To be sure that I understood his volte-face, when he had finished his talk I asked, "So you think that scrapie and BSE can arise spontaneously in animals when PrP^{C} refolds into PrP^{Sc}?"

"Yes, that's the only way to explain the Stetsonville outbreak," he replied. "There were no sheep within a hundred-mile radius of the mink farm where TME occurred, so you can't blame it on a virus. This is dairy country, and there are lots of old cows."

"Do you think the Stetsonville outbreak eliminates the possibility of a viral reservoir in sheep that periodically causes scrapie, and rarely BSE in cattle?" I asked, heart in mouth.

"I think it's difficult to reconcile TME in Stetsonville with the virus hypothesis. Your ideas about prions make the most sense, especially when you consider sporadic CJD, which occurs all over the world," he replied. "And look at BSE. The infectious form of prion disease has brought down more than fifteen thousand cattle this year alone, out of a U.K. cattle population of ten million."

Marsh and his graduate student Richard Bessen went on to study the properties of the Hyper and Drowsy prion strains. They found that the Hyper strain caused disease in hamsters about 65 days after inoculation and the Drowsy strain after about 170 days.[13] The Hyper strain was similar to the hamster scrapie strain that Marsh and Kimberlin had isolated two decades earlier, with respect to incubation time, neuropathologic changes, and molecular characteristics. The Drowsy strain, however, proved different from all other previously studied strains. It exhibited little resistance to digestion by proteases and its behavior in the centrifuge differed from that of other prions. While Hyper and other strains formed large aggregates that readily sedimented, the Drowsy strain did not form tight complexes, resulting in protein aggregates of many sizes.

The 170-day incubation period for the Drowsy strain in hamsters was like that of a strain first isolated by Richard Kimberlin and known as 139H, yet PrP^{Sc} from the 139H strain was highly resistant to digestion by proteases. This finding indicated that differences in the susceptibility of PrP^{Sc} to protease treatment were independent of the length of incubation time, the most distinctive feature of prion strains. This was hardly support for my hypothesis that prion strains represented different conformations of PrP^{Sc}. To the contrary, the comparison of Drowsy with 139H argued for another mechanism altogether.

What happened next was serendipity at its best. Pierluigi Gambetti sent me some of his samples of brain tissue from Italian victims of fatal familial insomnia, and Ruth Gabizon of Hadassah University Hospital in Jerusalem sent me samples of brain tissue from Libyan Jews who had died of familial CJD. Glenn Telling, a gentle, soft-spoken, superb postdoctoral fellow, injected the fatal familial insomnia and familial CJD samples into "humanized" Tg (transgenic) mice—mice that expressed a PrP gene that was part human and part mouse (figure 30). We found that the prions from the fatal familial insomnia patients caused PrP^{Sc} deposition only in the thalamus, just as Gambetti and Lugaresi had described in their patients (see chapter 12). In contrast, the prions from familial CJD patients provoked widespread deposition of PrP^{Sc} in the brain, just like that found in the brains of the Libyan Jews whose tissues Gabizon had sent me. Fatal

Figure 30 Led by Glenn Telling (above), in collaboration with Pierluigi Gambetti, Elio Lugaresi, and Ruth Gabizon, we showed the prion strain in the human brain inoculum determined the size of the PrP 27–30 in the brains of the Tg mice.

familial insomnia and familial CJD were caused by different mutations in the PrP genes of these patients.

The breakthrough came when Telling sent samples of the human and Tg mouse brains to Gambetti for analysis. He first converted the PrP[Sc] to PrP 27–30 by protease digestion and then removed the sugar chains attached to PrP 27–30. Using SDS gel electrophoresis, he measured the size of the PrP in the brain samples from fatal-familial-insomnia patients and Tg mice inoculated with fatal familial insomnia extracts. After removal of the sugar chains, the PrP 27–30 from the brains of the fatal familial insomnia patients and the inoculated Tg mice measured at 19 kDa (see chapter 8). When he repeated the experiment with the brain samples from the familial CJD patients and Tg mice, he found that the size of the PrP 27–30 was 21 kDa. Clearly, it was the inoculum that controlled the size of the PrP 27–30, by determining the regions of PrP[Sc] accessible to protease digestion. The overall shape of the protein dictated which segments of the amino acid chain were susceptible to cleavage by proteases.[14] In the patients with these inherited prion diseases, different mutations had generated two different shapes of PrP^{Sc}, but the PrP gene in the Tg mice hadn't mutated; rather, the shape of PrP^{Sc} in the human-brain inocula had been imposed on the PrP^C in the mice, acting as a template in the production of PrP^{Sc}. We now had unambiguous evidence that strain-specific properties were enciphered in the shape of PrP^{Sc}. However, the mechanism of protein templating remains obscure—as does the molecular language through which incubation times and patterns of PrP^{Sc} deposition in the brain are specified.

Another line of evidence emerged from a study conducted by a superb neurologist postdoc James Mastrianni on an American patient with fatal insomnia, whom he examined at UCSF. After the patient died, Steve DeArmond removed the brain and found lesions in the thalamus similar to those reported by Lugaresi and Gambetti in the Italian patients with fatal familial insomnia. Jim sequenced the PrP gene of the patient but did not find the expected fatal-familial-insomnia mutation at codon 178. This was apparently a spontaneous case of fatal insomnia. "What are we going to call *this* disease?" I mused. "Sporadic fatal insomnia!" Jim replied.

"If that's the right name," I said, "then Gambetti and Lugaresi should reverse the order of 'fatal' and 'familial.' They have defined FFI as fatal familial insomnia, not familial fatal insomnia."

But there was, of course, rather more to Jim's discovery. When he measured the size of the PrP 27–30, he found that it was 19 kDa. This meant that the patient's PrP^C had spontaneously folded into PrP^{Sc} with a shape that caused fatal insomnia—rather than having been inherited that way. And when he transmitted the prions from his deceased patient's brain to Tg mice, he found that the size of PrP 27–30 in the mouse brains was 19 kDa.[15] Jim's findings showed that the fatal insomnia trait (or phenotype, as geneticists call it) is enciphered in the shape of PrP^{Sc} and independent of the amino acid sequence—that is, the PrP mutation at codon 178 found in fatal-familial-insomnia patients is not required for fatal insomnia to develop. All that is needed is for PrP to fold into the PrP^{Sc} conformation that enciphers the insomnia phenotype.

More evidence followed when we examined other strains of prions. As difficult as it was to believe, the data argued that biological information is carried in the shape of PrP^{Sc}, which dictates both the incubation time and the pattern of neuropathology. Instead of just two different biologically active conformations of a single protein—namely, PrP^C and

PrP^{Sc}—we were now finding many. Clearly, PrP^{Sc} could adopt a multitude of shapes. Its remarkable plasticity was surprising but offered an explanation for the existence of different "strains." The last real argument against the prion—the existence of different prion "strains"—was no longer a conceptual obstacle. No longer did scientists need to invoke a nucleic acid as an explanation for its diversity.

While these studies were in progress, Ana Serban, a careful and dedicated research associate, performed a series of experiments designed by Jim Cleaver, an imaginative professor of radiobiology. She irradiated two strains of prions with increasing doses of ultraviolet light. Both strains proved equally resistant to inactivation, yet another indication that the difference between them was not due to a nucleic acid. Cleaver had calculated that based on the prion's resistance to ultraviolet light, a putative nucleic acid inside it could contain no more than five bases.[16] Since three bases are required to encode one amino acid, a hypothetical prion genome of five bases could encode only a single amino acid. Small proteins contain more than one hundred amino acids, and some proteins are composed of several thousand.

A new twist on the search for that mythical nucleic acid emerged in the fall of 1996, when I visited Richard Setlow, the associate director of Life Sciences at the Brookhaven National Laboratory on Long Island and a renowned expert on genetic diseases and DNA repair. He showed me a startling piece of data he had published in 1957. It was a graph showing inactivation spectra for two proteins, the enzymes aldolase and trypsin, using various wavelengths of ultraviolet light (figure 31).[17] Then he showed me what happened when he superimposed the data that Tikvah Alper and Raymond Latarjet published in 1970 (see figure 6). Their data for the "scrapie agent" fell precisely on the trypsin curve. The fit was nearly perfect. "Wow!" I exclaimed. "I'm speechless! I spent years—or at least countless hours—trying to understand their data, and you had the answer in a paper you published thirteen years before they published theirs. Why didn't Alper and Latarjet use your published data? They should have known your published work—they must have forgotten it. If I'd been smart enough to find your paper, I would have known that the scrapie agent was a protein and the problem would have been solved. End of story!"

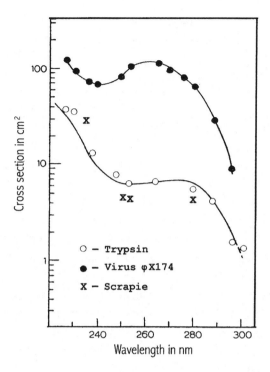

Figure 31   During a visit to the Brookhaven National Laboratory on Long Island, Richard Setlow showed me a startling piece of data he had published in 1957: a graph delineating different inactivation spectra for two proteins—the enzymes aldolase and trypsin—using various wavelengths of UV light. Then he showed me that the data for the scrapie agent reported in 1970 by Tikvah Alper and Raymond Latarjet: their data points (x) fell precisely on the trypsin curve.

"Well, that's one way to look at it, but you still needed to do all the protein chemistry, molecular biology, and genetics you did," said Setlow. "Without those studies, no one would have believed that prions exist."

Interlude: The Key to Interpreting the Trypsin Inactivation Curve

Proteins absorb ultraviolet light maximally at 280 nanometers, due to their aromatic amino acids, such as tyrosine and tryptophan, which absorb light at 280 nm. But the absorption spectrum doesn't always correspond to the inactivation profile. When a protein has cysteine residues, the absorption and inactivation spectra are different. That's the case with both trypsin and the prion protein. The absorption spectrum of a protein has a maximum at 280 nm, due to the tyrosine and trytophan residues, and a minimum at 260 nm. But the inactivation of a protein with cysteines can be the same at 260 nm and at 280 nm. When photons at 260 nm strike cysteine residues, they damage the protein more effectively than when they collide with aromatic amino acids, like tyrosine and tryptophan. In every mammalian PrP, two cysteines have been found.

On the flight back to California, I mused about what I had learned from Setlow. Had I known about the trypsin results, had I understood the effects of cysteine in ultraviolet light studies, had I understood the difference between absorption and inactivation spectra, might my investigation of the scrapie agent have progressed more rapidly? I felt he was probably right, in that all the hard work would have had to be done anyway.

What we knew by now was that prions were infectious proteins and that the properties of different strains were enciphered in its various conformations. This seemed like a satisfying accomplishment—but we still did not understand the language of prions. Later, we would learn that less stable conformations of PrP^{Sc} generally encipher shorter incubation times while more stable ones produce longer times. These findings had to wait until we were able to create synthetic prions from PrP produced in bacteria using recombinant DNA technology.[18] Despite these advances, we have been unable to decipher how prion strains dictate variations in the neuropathology.

16

Stockholm

A s the concept of infectious proteins became accepted, more and more people, both inside and outside science, began speculating about whether (or when) there would be a Nobel Prize for the work. I was never sure how to handle such comments. I viewed them as idle chatter, given all the negative press I'd engendered. The very people who had been contemptuous of the prion concept were beginning to realize that their scorn had served only to elevate the achievements of my colleagues and me. But others still did not understand the evidence, and their continued skepticism burnished my image as a daring scientific rebel. A little more opprobrium and I would start looking like a visionary!

After I received the Lasker Award in September 1994, talk about a Nobel Prize for the discovery of prions became commonplace. While many Lasker laureates were later recipients of Nobel Prizes, I could not recall any instance of someone receiving the two awards in the same year.[1] With the announcement of the Nobel Prize in Physiology or Medicine in October 1994 going to Alfred Gilman of the University of Texas and Martin Rodbell of the National Institute of Environmental Health Sciences for their discovery of G-proteins, which enable cells to communicate with one another, my only regret was that I would not get a permanent assigned parking place at UCSF. Finding a place to park my car at UCSF wasted almost fifteen minutes every morning and another ten minutes in the evening. That was twenty-five minutes I could have spent with my family or reading a good book. The following spring, when I encountered Gilman at the annual meeting of the National Academy of Sciences, he described the middle-of-the-night telephone call informing him of the Nobel. "Someday you'll get a call like that," he said.

"That happens to other people, not me."

Gilman laughed. "I used to get so annoyed at people who said, 'Someday you're going to win a Nobel Prize.' Let me tell you, the best thing about winning a Nobel Prize is that no one will ever say to you again, 'Someday you're going to win a Nobel Prize.'"

A few months later, the gods granted me a parking place. While I was at lunch with one of the UCSF vice chancellors, he handed me a letter from Joseph B. Martin, who had recently become the chancellor after Julie Krevans retired. On the occasion of the fiftieth anniversary of the Lasker Awards, Martin had decided to provide free parking to all UCSF past winners. I thought this was nice, but I immediately began to think about how to convert it into an actual assigned parking place. Should I wait a day, two days, a week, before broaching the subject? As it happened, I waited not ten minutes. "This is lovely," I told the vice chancellor, "but is there some way I can continue to pay for the parking and convert this privilege into an assigned spot?" An assigned spot came along with the perk, he said. I was in heaven!

When fall 1995 and then 1996 came and there was still no call from Stockholm, I was not disappointed: From my assigned parking place to my lab was only a three-minute walk. Although I had to endure comments like "Someday you're going to win a Nobel Prize," I had learned to ignore them. I had to. The problem was that each year when I did *not* get a call from Stockholm, there was the unstated assumption that there was something wrong with me—something horribly lacking. I began imagining spending thirty or forty years waiting for a Nobel and never receiving one. I understand that this is not an uncommon preoccupation among scientists at a certain level of accomplishment.

I knew that several groups had nominated me, but apparently in vain. My work would not get better—our discovery of prions had occurred in the early 1980s. And the media had not lost its appetite for attacking me. A book titled *Deadly Feasts* by Richard Rhodes appeared in spring 1997. Rhodes had written to me almost a year earlier asking for an interview, which I refused despite his reputation as the author of well-regarded books about the making of the atomic and hydrogen bombs. As usual, I thought that little or no good could come from talking to journalists. When Rhodes's new book was published, nearly 80 percent was devoted to the work of Carleton Gajdusek in the highlands of New Guinea, and about 5 percent essentially trashed me as a second-rater.

In September 1997, I flew to Buenos Aires to give three lectures at the Sixteenth International Congress of Neurology and another lecture at the University of Buenos Aires. The day after my first talk at the congress, the daily newsletter had my picture on the front page with the caption "Stanley Prusiner, future Nobel Prize winner, delivers lecture." Not even in South America could I escape. Gilman's prophecy continued to haunt me.

When I left Buenos Aires, I flew to Stockholm, where Rudolf Riegler, a biophysicist at the Karolinska Institute, had organized a symposium on prions under the auspices of the Swedish Academy of Sciences. The next day, I met Riegler and two colleagues, Hans Jornvall and Arne Holmgren, for lunch. The three Swedish scientists all studied protein structure, but each employed a different technique: Jornvall used X-ray crystallography, Holmgren nuclear magnetic resonance spectroscopy, and Riegler fluorescence spectroscopy. I was impressed by how knowledgeable they were about prions; they were well up on the literature and better informed than many people working in the field.

On Tuesday morning, a large crowd assembled at the Swedish Academy, where my friend Erling Norrby was now the executive secretary. A few minutes before the session began, Riegler turned to me and said, "You'll be the first speaker. I'm giving you fifty minutes for your lecture and the others will have forty." Seeing my surprise, he explained, "You clearly started the prion work, so you should get a little more time." Riegler's kind instructions put me at ease, and with this opportunity to make my presentation clearer, more deliberate, and unhurried, I delivered one of my best lectures.

When I finished the talk, my erstwhile collaborator Charles Weissmann took the podium. About two-thirds of the way through his lecture, he returned to the joint work we had published five years earlier. He described knocking out the PrP gene only to find that the mice seemed quite normal except for some subtle defects in sleep-wake cycles. He hypothesized that since PrP was not expressed throughout the lifetime of the knockout mice, the mice might adapt to its absence. To test his conjecture, Weissmann suggested using a system developed by Herman Bujard in Heidelberg, wherein an artificial PrP gene could be introduced into the mouse in which the PrP had already been knocked out. The knockout mice would express PrP, but it could be turned off in adulthood by feeding them the antibiotic doxycycline. In this scenario, the acute shutdown of PrP expression in an adult animal might produce a recognizable disease state. Weissmann, however, had no data. He had spent almost half his lecture ruminating on a future experiment.

When Weissmann finished, his talk and mine were open for questions. Because most of the queries were directed to me, Weissmann sat down after a few minutes, and I dealt with the remaining questions—all the while pondering what to do. More than three years earlier, I had asked Herman Bujard to help us produce mice with an inducible PrP gene whose expression could be controlled by doxycycline. Zeev Meiner, a young neurologist from Hadassah University Hospital in Jerusalem, started this work at UCSF as a postdoctoral fellow; after he returned to Israel, Patrick Tremblay, a postdoc from the Max Planck Institute in Göttingen, carried it forward. We had already completed the experiment that Weissmann had proposed in his talk, and it had proved of no value in illuminating the function of normal PrP.[2] Should I disclose our results publicly or privately? The more gentlemanly way was to keep this news to myself and tell Weissmann at the break, but I decided I had to describe our results publicly then and there. For one thing, the idea was a bust; for another, if I revealed the data later, some people unfamiliar with the prolonged time frame of such studies might have thought I had simply executed Weissmann's wonderful idea. Looking directly at Weissmann, I said, "Charles, I want to tell you the results of the experiment you outlined at the end of your talk. We produced mice carrying inducible PrP genes and fed them doxycycline after they reached adulthood to shut off PrP expression acutely. After two months on doxycycline, we have seen no sign of illness. Thus the acute repression of PrP gene expression in adult mice has not given us any insight into the function of the protein."

Weissmann seemed stunned. At the break, he said to me, "I guess you call that real one-upmanship."

"Your presentation left me no choice," I replied. Although I felt sorry for him, I didn't at all mind reporting on our efforts to elucidate the function of PrP.

When the program resumed, Detlev Riesner described some of the studies we had done together during our long joint search for a nucleic acid in purified prion preparations and our studies on the unfolding and refolding of PrP 27–30.

Next up was Kurt Wüthrich, of the Eidgenössische Technische Hochschule in Zürich, who spoke about his nuclear magnetic resonance work on the structure of proteins (it was to garner him a Nobel Prize five years later). With the first phase of our own nuclear magnetic resonance studies completed nearly six months earlier, I felt much more at ease when Wüthrich spoke than I would have if we lacked our own structural data. Wüthrich began his talk by discussing one of three hypotheses set forth by the British mathematician John S. Griffith in a short letter to *Nature* in 1967.[3] Based on Tikvah Alper's results, Griffith had offered the prescient speculation that the scrapie agent is a protein that adopts an abnormal conformation. I would have been surprised if anyone had taken this idea seriously at the time; Chris Anfinsen's work showing that the amino acid sequence of RNase specifies a single biologically active conformation was the prevailing dogma. When I first read Griffith's paper, five years after it was published, I wondered whether he had ever heard of Anfinsen's seminal studies. I assumed he did not know the literature on protein folding, and I dismissed his musings as unimportant. Griffith had sandwiched his conformational hypothesis between two other posits, one of which proposed that the scrapie agent was a protein encoded by a cellular gene that was expressed only during infection and the other that the scrapie agent was like an antibody that both reacted with itself and stimulated more of its own production.

About twenty minutes into his talk, Wüthrich gave a demonstration of his confrontational nature. He showed a slide with the three-dimensional model of PrP^C that Fred Cohen and I had proposed, but he had modified the model by placing a giant black "X" across it with electrical tape. Loud gasps of embarrassment came from all corners of the auditorium at this expression of overt hostility. Seemingly very angry, Wüthrich bellowed, "Fred Cohen's model is wrong!" When his talk concluded, we adjourned for lunch, and I asked him for the slide, with the idea that I might someday publish it or a facsimile, but he cannily refused to give it to me, though he said he would be happy to explain how I could construct my own, using our published model and black electrical tape, which is what I did.

Before returning to California, I flew to Düsseldorf with Detlev Riesner (see figure 24). After spending several hours working on a manuscript together, we went to dinner, where we reviewed the events of the previous days.

"What do you think about your chances for a Nobel Prize?" he asked, rather circumspectly.

"I don't know." A few seconds passed, and I said sotto voce, not wanting anyone to overhear, "I think my chances are about fifty-fifty, based primarily on the extraordinary knowledge of prions that Riegler, Jornvall, and Holmgren displayed at lunch the day before the symposium—but please don't repeat this."

The Phone Call

The announcement of the award was to be made in less than two weeks. Back home, I flew to Washington on Sunday morning, October 5, at the behest of the Food and Drug Administration to participate in a spongiform encephalopathy advisory panel meeting the next day. After checking into the Holiday Inn in Bethesda, I eventually fell asleep around midnight. At 5:05 A.M. EDT, I was awakened by a call from Nils Ringertz, acting secretary of the Karolinska Institute's Medical Nobel Assembly, who informed me that the assembly had just finished voting to give me the 1997 Nobel Prize in Physiology or Medicine. Despite all my hopeful feelings, despite the encouraging Stockholm meeting, I was flabbergasted.

"I want you to know," Ringertz continued, "that the Nobel Assembly voted to award the *entire prize* to you."

Now I was dumbstruck. Ignoring the silence on my end, Ringertz told me that the prize was being awarded for "the discovery of prions, a new biological principle of infection." He asked if I agreed with the wording. It flashed through my mind that "disease" might be a better word than "infection," because prions not only infect but also produce genetic and sporadic neurodegenerative illnesses. As quickly as this thought appeared, I decided to let it go. They had spent more time thinking about the wording than I had. Moreover, what if I seemed too argumentative? They still had time to change their minds!

"Sometimes," Ringertz continued, into the void, "when I have called, people have thought it was not I. Do you remember meeting me?"

"Yes, we had dinner together at Erik's Restaurant in the Old Town of Stockholm in November 1996," I replied.

"That's right," he answered. "I hope you can arrange your schedule so that you and your family can come to Stockholm in early December for the Nobel Prize ceremony."

"I'm sure we can manage." I thanked him and hung up. Certainly my wife Sandy and our two daughters, Helen and Leah, could find a way to schedule a trip to Stockholm. I sat there in a daze for ten minutes or so until Sandy called me. She had learned five minutes before I did, since Ringertz had of course called our home first. After we talked, I tried to go back to sleep. Ringertz had asked me not to tell anyone for twenty-five minutes, until he announced it to the press. I said that was fine with me.

At 5:30 A.M., I turned on CNN to confirm that the call was real—to be sure it was not a prank. But there was no news from Stockholm—just a recital of the latest mayhem and the foreign stock market tallies. Only then did I begin to think seriously that this could all be a ruse. When Ringertz asked whether I remembered him, I should have asked him where we met, instead of recalling that we had had dinner together at Erik's. At 6:00 A.M. and 6:30 A.M., still no news from Stockholm. What was going on? At 7:00 A.M., I decided to turn on the radio and tuned it to an ABC affiliate in the Washington, D.C., area, where I heard a rich baritone voice announce, "An American, Dr. Stanley Prusiner, has won this year's Nobel Prize in Physiology or Medicine." *Phew!*

At a news conference announcing the prize in Stockholm, Ralf Pettersson, the vice chairman of the Nobel Committee at the Karolinska, stated, "During the whole of the 1980s, the prion was very controversial. Acceptance took a while. . . . We now have a theoretical basis for designing new drugs to prevent the transformation to a pathological form of protein. It will help in research to understand dementia in other forms, like Alzheimer's, which have many features in common." After the announcement, my emotions consumed me—joy, of course, accompanied by an inexplicable exhaustion. And almost immediately I began hearing from friends all over the planet. Among the hundreds of notes and e-mails were beautiful letters from Nobel laureates, one of whom told me that the award would be accompanied by "the party of your lifetime"—an intriguing description that proved correct. To my delight, I would learn that the Swedes had perfected the celebration of intellectual achievement.

The Press Conference

On Wednesday, December 3, 1997, I arrived in Stockholm, and the next day I faced a legion of skeptical Stockholm newspaper reporters. The Nobel Foundation convened a press conference in a building called the Nobel Forum on the grounds of the somewhat drab Karolinska Institute. Most of the grass was covered with melting snow and mounds of dirtied, glistening ice along the edges of the streets and sidewalks. A red brick façade provided a rather unimposing exterior for the Nobel Forum that was interrupted by a few white wood-frame windows; in stark contrast, the interior exuded warmth and an almost hallowed decor. On the ground floor of the Nobel Forum building was a beautiful, solemn auditorium with oiled cherry trim and desks that held about a hundred people, while the upper floors contained precious papers documenting deliberations beginning with the first Nobel Prize in Physiology or Medicine in 1901.

As I entered the auditorium that morning at 9:00 A.M., I felt a blast of frigid arctic wind. Somehow, I could tell that this was not going to be a celebration of science. Rather, it was going to be another barbeque of Stan Prusiner—how could a scientist be so wrong? Even more disturbing, how could the Nobel Assembly at the Karolinska have made such a colossal error?

The day before, the major newspaper of Stockholm, the *Svenska Dagbladet,* carried a front-page story reporting doubts about the existence of prions. Maybe Stan Prusiner's discovery wasn't real? With more time and more hard work, perhaps a core of well-protected DNA or RNA would be found, demonstrating that prions were really viruses. If so, the Nobel Assembly had made a premature decision—a terrible mistake. Why couldn't they have waited just a few more years?

As the press conference began, a sick, tense feeling engulfed my lower abdomen. My stomach started to ache. What is wrong with these guys? Why don't they read my papers and understand that the data are solid? Why don't they acknowledge that others have independently reproduced many of my key findings? Where are their heads?

I began by explaining how I had come to the problem of what causes a rapidly progressive dementia in people in their fifties and sixties. The illness that I described is called Creutzfeldt-Jakob disease, often abbreviated CJD. Next, I told them about a similar disease in sheep called scrapie that could be studied in laboratory rodents. I explained that shifting my studies from mice to hamsters and redesigning the method of measurement accelerated my work almost eightyfold. And it was this change in my study procedures that led to my discovery of prions.

These descriptions were fine, but the press was primarily focused on whether there was even a remote chance that prions could be real rather than simply a figment of my imagination. They seemed unable to envision the possibility that viruses did not cause scrapie. After all, viruses are minute infectious agents, much smaller than bacteria—we've known about viruses for more than a hundred years. Why not a virus—a special, peculiar, atypical virus, but a virus nevertheless? What is this nonsense about infectious proteins, or prions, as I called them, being responsible for diseases thought to be viral illnesses?

Trying not to show my impatience, I explained that I initially thought viruses caused kuru, CJD, and scrapie, but that years of searching for the putative virus had led me to conclude otherwise. I recited how I had enriched brain fractions for scrapie infectivity and removed more than 99 percent of the impurities. This procedure revealed the unusual properties of the scrapie agent, while the evidence for a virus steadily faded.

Some reporters insisted that I must have made a frightful mistake. Neither my findings nor my interpretation could be correct. I just had to be wrong—any reasonable person knew that prions were impossible. Every road sign, every clue, every signal pointed to a virus. How could the Nobel Assembly be so reckless and irresponsible to award the 1997 Nobel Prize in Medicine to me? And the Assembly highlighted its brash and foolish actions by awarding the entire prize to me. I was the sole recipient.

So there I stood, puzzling whether the press expected me to impale myself on a sword of guilt and sorrow for the negligent actions of a group of fifty professors, who had been elected to the Nobel Assembly by their peers. The Assembly had been charged with guarding the sacred, consecrated ground of scientific accomplishment for almost a hundred years, but this time it had clearly failed! Its record of past successes in recognizing the outstanding achievements of scientists all over the world was now tarnished forever. Its reckless decision was far more than a blemish, it was a seismic blunder.

After a two-hour inquisition, Niels Ringertz, the secretary of the Nobel Committee, declared the press conference closed. The Nobel Committee, consisting of a half-dozen Karolinska professors, had recommended to the Assembly that I be the sole recipient of the 1997 Prize. Once the reporters exited, Ringertz apologized for my having been subjected to such intense questioning from a large group of dubious reporters. I responded by telling him that I was not a neophyte in the land of prion naysayers, and I described my policy of refusing to speak to the press for more than a decade.

Triumph of Experimental Science

It was a time to hail the triumph of experimental science in rolling back the frontiers of our ignorance, and a time, too, to remember how much my experience working in Stockholm with Olle Lindberg thirty years earlier had meant to me. It was a special time for my family and friends, who gathered in Stockholm. Besides Sandy and our two daughters, my father and brother made the trip. Numerous other friends and supporters came, including Oded and Henia Abramsky, Hank and Annik Baron, Walter and Connie Burke, George and Fran Carlson, Fred and Carolyn Cohen, Steve and Bernadette DeArmond, Bob Fishman, Ruth and Alberto Gabizon, Don and Audrey Gilden, Darlene and Don Groth, Steve Hauser, Julie and Patsy Krevans, Mac and Marj McCarty, Barbro and Barney Osher, Detlev and Hannalore Riesner, Fred Seitz, Albert and Anat Taraboulos, and Charley and Edna Yanofsky. It was also a time to see many of my Swedish friends and colleagues, some of whom I had first met when I was a medical student working on brown fat. These included Barbara Cannon, Gustav Dallner, Lars Ernster, Tomas Hökfelt, Krister Kristensson, Olle Lindberg, Erik Lycke, and Erling and Margaretta Norrby.

With the announcement of the prize, I was confident that the controversy over prions had finally ended. I was naïve. Richard Rhodes (was he still smarting from my rejection of him while he was writing his Gajdusek opus?) proceeded to publish an article in the *New Yorker*—entitled "Pathological Science" and subheaded "Stanley Prusiner is being awarded a Nobel Prize for his groundbreaking theory about mad-cow disease. But what if he's wrong?"—in which he predicted that the virus we had missed would soon be found and claimed that the Swedes had made a terrible mistake:

> Prusiner's contribution to medical science has been sensational. In 1982, he posited a new kind of infectious agent, unique in nature, to explain the unusual characteristics of a group of rare, transmissible spongiform encephalopathies (T.S.E.s)—diseases that kill by damaging the brain with knots of junk protein and tiny holes. In the decade that followed, as he worked to prove his theory, a new previously unknown TSE emerged in an epidemic that panicked millions. The . . . outbreak in England of bovine spongiform encephalopathy (B.S.E.), commonly known as mad-cow disease, resulted in the slaughter of nearly two million animals. When young people . . . began dying as well [of new variant Creutzfeldt-Jakob disease], probably infected by eating tainted meat, fears arose of a similar epidemic in human beings. This climate of urgency focussed scientific and public attention on Prusiner's controversial theory that the infectious agent for the T.S.E.s is not a virus or a bacterium but a protein devoid of the nucleic acids DNA and RNA. In Prusiner's coinage, the diseases were caused by a "prion"—a "proteinaceous infectious particle."
>
> . . . Despite Prusiner's insistence that the puzzle has been solved, the last word has not yet been written about the prion's role in causing such diseases. In fact, by awarding the prize to Prusiner this year the fifty-three Karolinska scientists who make up the Medical Nobel Assembly explicitly took his side in the ongoing debate—a debate that may have profound consequences for future research. . . . The Karolinska neurologist Lars Edstrom was dismissive of the skeptics when he spoke to the *Times* upon announcement of the prize. "There are still people who don't believe that a protein can cause these diseases, but we believe it. From our point of view, there is no doubt."
>
> The tone of these endorsements offends many researchers in the field, including the neuropathologist Laura Manuelidis. . . . Although some of her colleagues cede more ground to Prusiner, by her count

only four of the fourteen major T.S.E. research labs . . . espouse prion theory. . . . At a conference in Paris in 1995, Manuelidis told T.S.E. researchers that, despite the prion bandwagon, almost everything about the T.S.E.s points to a "slow virus" (so called because of the long incubation periods), with PrP acting as a gatekeeper at some stage in the organism's life cycle. . . .

Taken together, Prusiner's experiments demonstrate at least that PrP is crucial to T.S.E. infection. Yet if protein alone is responsible for T.S.E., people at different laboratories told me, then it ought to be possible to synthesize PrP in a test tube and cause infection. That's the Holy Grail of prion research. But it's been tried, so far without success. Nevertheless, Prusiner believes that with more research it's only a matter of time. . . . Prusiner's work, for all its energy and ambition, may yet turn out to be an example of what Nobel-laureate chemist Irving Langmuir once called "pathological science" . . . sensational scientific "discoveries" that turned out to be false—not fraudulent but merely self-deluded.[4]

And of course Gary Taubes weighed in with an equally vitriolic analysis, in an article entitled "Nobel Gas," which began:

Sure, Stanley Prusiner deserves a prize—for his persistence, not for his prions. Nobody said the Nobel Committee was infallible. It did, after all, give Henry Kissinger the peace prize in 1973. But the folks in Stockholm have traditionally been conservative about whom they bestow scientific awards upon. . . . If it turns out that viruses *do* cause the diseases [Mad Cow and Creutzfeldt-Jakob], then Prusiner will have won the prize for the discovery of something spectacularly wrong.

Good science, not just Nobel Prize–caliber science, depends on hypothesis and test, and then the rigorous demonstration that the preferred interpretation of the data was the only interpretation. In other words, remarkable results demand remarkable evidence. In the case of Prusiner's prize, the Nobel Committee has settled for enthusiasm and single-mindedness.[5]

An unanticipated perspective on Taubes emerged the following spring, during a conference organized by the Nobel Foundation at an old monastery in the archipelago east of Stockholm. As the meeting broke for lunch, the genome pioneer J. Craig Venter came up to me and said, "I have an amusing story for you. Gary Taubes came to my office a couple of months ago and tried to convince me to let him write my biography. I said, 'Didn't you write a terrible piece about my friend Stan Prusiner some years ago?' Taubes looked uncomfortable for about thirty seconds, and then a big smile came over his face and he said, 'My article got Stan Prusiner a Nobel Prize.'"

I burst out laughing. "That's the definition of chutzpah," I said.

"I told him I thought the *Discover* article about you was pretty terrible," Venter continued, "and I asked him to send me a copy of it and copies of all his other articles and books about scientists before I decided whether I'd help him write my biography." Taubes never responded to Venter's request, and Venter ended up writing his autobiography.[6]

The Celebration

The next evening at a small dinner given by the Nobel Committee, I had the opportunity to meet all of the scientists and their spouses who had been my advocates. About twenty people attended the dinner, which was held in an elegant room at the Grand Hôtel, where all the laureates and their families were staying. The tables and walls were covered with fabrics in various festive shades of gold. Gösta Gahrton, a professor in the Karolinska Institute's Department of Medicine and chairman of the Nobel Committee, took me aside and said, "We love prizes like this one, where the path of discovery is so clear that we could award it to one person. We know all about the many obstacles you encountered along the way, and we hope that the stature that comes with the Nobel Prize will help with your funding problems. We're aware of the difficulties you have encountered in finding adequate support for your research. Your perseverance has been remarkable."

In the afternoon of the fourth day in Stockholm—December 10, the anniversary of Alfred Nobel's death—the laureates and their spouses assembled in the lobby of the Grand. From there, we drove to the Concert Hall in a motorcade of black Volvo limousines. My excitement was building. How could this be happening to me?

In the Concert Hall, which was packed, my fellow laureates and I were seated on the stage with members of the Swedish academies. After each of the physics and chemistry laureates received his prize, my turn came. Ralf Pettersson was called to the podium and asked to present me to King Carl Gustaf. Pettersson began,

Your Majesties, Your Royal Highnesses, ladies and gentlemen,

This year's Nobel Prize in Physiology or Medicine has been awarded to Stanley B. Prusiner for his discovery of prions—a new biological principle of infection. What is a prion? It is a small infectious protein capable of causing fatal dementia-like diseases in man and animals. It has been known for approximately a century that infectious diseases can be caused by bacteria, viruses, fungi, and parasites. All these infectious agents possess a genome, the hereditary material that provides the basis for their replication. The ability to replicate is essential for the manifestation of the diseases they cause. The most remarkable feature of prions is that they are able to replicate themselves without possessing a genome; prions lack hereditary material. Until prions were discovered, duplication without a genome was considered impossible. This discovery was unexpected and provoked controversy.

. . . Prusiner set out to purify the infectious agent, and after ten years of hard work he obtained a pure preparation. To his great surprise, he found that the agent consisted only of a protein, which he named "prion," a term derived from "proteinaceous infectious particle." Strangely enough, he found that the protein was present in equal amounts in the brains of both diseased and healthy individuals. This discovery was confusing, and it was generally concluded that Prusiner must have arrived at the wrong conclusion. How could a protein cause disease if it was present both in diseased and healthy brains? The answer to this question came when Prusiner showed that the prion protein from diseased brains had a completely different three-dimensional conformation. This led Prusiner to propose a hypothesis for how a normal protein could become a disease-causing agent by changing its conformation. . . . But how can a protein replicate without a genome? Stanley Prusiner suggested that the harmful prion protein could replicate by forcing the normal protein to adopt the shape of the harmful protein in a chain reaction–like process. . . .

The hypothesis that prions are able to replicate without a genome and to cause disease violated all conventional conceptions and during the 1980s was severely criticized. For more than ten years, Stanley Prusiner fought an uneven battle against overwhelming opposition. Research during the 1990s has, however, rendered strong support for the correctness of Prusiner's prion hypothesis. The mystery behind scrapie, kuru, and mad cow disease has finally been unraveled. Additionally, the discovery of prions has opened up new avenues to better understand the pathogenesis of other more common dementias, such as Alzheimer's disease.

Stanley Prusiner, your discovery of the prions has established a novel principle of infection and opened up a new and exciting area in medical research. On behalf of the Nobel Assembly at the Karolinska Institute, I wish to convey to you my warmest congratulations, and I now ask you to step forward to receive your Nobel Prize from the hands of His Majesty the King.

Pettersson's speech (which was in Swedish) had seemed to go on forever. I was growing increasingly anxious, even though, like the other recipients, I would not have to speak at the awards ceremony. Inexplicably, I was worried about walking to the middle of the stage to shake hands with the king. At that time, he would present me with a magnificent certificate in a red leather folio embossed with gold and a small red leather box containing a gold medal with Alfred Nobel in profile. What if I tripped? What if I dropped the medal, as I went to shake the king's hand? Or what if I forgot the sequence of bowing to the king, turning left to bow to the queen, and then turning right to bow to the audience? My brain was filled with this *mishegas* when it should have been filled with only *mazltof* and joy. When Pettersson finally summoned me, I stood up with my stomach full of butterflies and advanced toward the king, who seemed preternaturally calm. I managed to shake his hand (figure 32) and hold onto the certificate and medal while bowing to him, the queen, and the audience in the prescribed order, then returned to my seat overcome with relief.

After the bestowing of certificates and medals, the Volvo motorcade carried us to the City Hall, where the Nobel banquet was about to begin. Unfortunately, the City Hall could not accommodate all the people who attended the presentation ceremony, so only a few of my friends and supporters, out of the fifty-five who had come to Stockholm, could enjoy the

Figure 32   Receiving the Nobel medal and certificate from the king of Sweden, Carl XVI Gustaf.

official banquet. Tickets for both the ceremony and the banquet were difficult to obtain. Requests from friends were invariably accompanied by an unspoken mantra: "If you're smart enough to win a Nobel Prize, you're smart enough to get one more ticket." Loyalty, a desire to be responsive, and the magnanimity inspired by good fortune fueled my dogged quest for tickets, which some people saw as excessive ("Stan's being pushy again!"). The irony was that when I was in Stockholm in 1967 the Nobel Foundation couldn't *give* the tickets away: It had to recruit Karolinska medical students and provide them with rented tuxedos. And now the ceremony-cum-banquet was one of the hottest winter attractions in all of Scandinavia.

For thirty-three ticketless friends, I rented a small ballroom at the Grand Hôtel, with a huge TV screen on which to view the proceedings. Afterward, they were treated to a satellite banquet, which was a modern replica of the first Nobel banquet, in 1901 (the first Nobel Prize that night was awarded to Wilhem Roentgen for his discovery of X-rays): hors d'oeuvres followed by filet of brill in white wine sauce served with prawns and mussels; a second course of fillet of beef with truffle sauce, goose liver, and asparagus; and a third course of breast of hazel grouse with salad, accompanied by a cream sauce and black currant jelly. The aptly named ice cream dessert, Succès Grand Hôtel, came with biscuits and petits fours. The wines served in 1901 were Niersteiner 1897, Château Abbé Gorsse 1881, and Champagne Crème de Bouzy. Regrettably, these had to be updated: Mâcon Blanc-Villages 1996, Bourgogne Château Laroche Bel Air Cuvée Grand Hôtel 1976, and a Lenz Moser Brut. I longed to be in both places—the City Hall and the Grand Hôtel at the same time—at once. Many of the satellite diners had contributed significantly to the work I was being honored for.

Meanwhile, the banquet in the City Hall began with a procession of waiters skillfully bearing covered plates down a grand two-story stone staircase to the tables, where, most impressively, they lifted almost thirteen hundred plate covers in unison. The first course consisted of a Jerusalem artichoke galette accompanied by smoked Landskrona salmon and lobster; the second of breast of wild squab and porcini ragout with potato and onion compote in a sweet-and-sour raspberry vinegar sauce. The wines accompanying this elegant cuisine were Champagne Pommery 1991, Brut Millésimé Crozes-Hermitage Domaine de Thalabert 1990, and

Paul Jaboulet Aîné Château Pajzos 1993. With the arrival of the dessert course (parfait of elderflower ice cream and strawberry sorbet) came another anxiety-provoking task: the short banquet speech. In preparation for delivering mine, I had read several Nobel banquet speeches, the most eloquent being those of William Faulkner, Richard Feynman, and Saul Bellow, but they had inspired me with nothing but self-doubt. After a perfectly respectable offering by Dario Fo, the Italian satirist, playwright, theater director, actor, and composer "who emulates the jesters of the Middle Ages in scourging authority and upholding the dignity of the downtrodden,"[7] it was my turn. As I stepped to the podium, my anxiety level soared, but somehow I managed to compose myself. I was hoping that I would not need my notes, which I had edited for the umpteenth time during the banquet, but that was not to be.

King Carl Gustaf, Queen Silvia, distinguished guests,

Tonight's awarding of the Nobel Prizes and this elegant banquet are a grand celebration of science and culture. These events honor the vision, courage, and wisdom of Alfred Nobel, a superb scientist in his own right. And they honor the Swedish people's uncommon sense of commitment to fulfill the enlightened wishes of Alfred Nobel; the Nobel Prizes have truly become important benchmarks in the history of science.

But the Nobel Prizes are much more than awards to scholars, they are a celebration of civilization, of mankind, and of what makes humans unique—that is, their intellect, from which springs creativity. . . .

People often ask me why I persisted in doing research on a subject that was so controversial. I frequently respond by telling them that only a few scientists are granted the great fortune to pursue topics that are so new and different that only a small number of people can grasp the meaning of such discoveries initially. I am one of those genuinely lucky scientists who was handed a special opportunity to work on such a problem—that of prions.

Because our results were so novel, my colleagues and I had great difficulty convincing other scientists of the veracity of our findings and communicating to lay people the importance of work that seemed so esoteric! As more and more compelling data accumulated, many scientists became convinced. But it was the "mad cow" epidemic in

Britain and the likely transmission of bovine prions to humans producing a fatal brain illness called Creutzfeldt-Jakob disease that introduced prions to the public. Yet the principles of prion biology are still so new that some scientists and most laymen, including the press, still have considerable difficulty grasping the most fundamental concepts.

... No matter how new and revolutionary the findings may be, as data accumulate, even the skeptical scholars eventually become convinced except for a few who will always remain resistant. Indeed, the story of prions is truly an odyssey that has taken us from heresy to orthodoxy.

Lastly, we celebrate on the occasion of these Nobel Prizes, the triumph of science over prejudice. The wondrous tools of modern science allowed my colleagues and me to demonstrate that prions exist and that they are responsible for an entirely new principle of infection.

Tack så mycket!

17

The Third Judgment of Paris

I n the 1860s, an imaginative group of mostly French painters began a revolution in their Parisian studios. These Impressionists, as they were later called, were subjected to years of harsh judgment and ignorant ridicule by the French artistic establishment. Impressionism would profoundly change art and artists forever. In 1874, the Impressionists held their first exhibition in Paris on their way to achieving worldwide acclaim.

One hundred and two years later, a second Judgment of Paris would be offered by a group of French oenophiles, who ranked both a cabernet and a chardonnay from California as number 1 over some exalted French wines. This surprising coronation of California wines would forever change global viticulture.

In late February 2012, Yves Christen of the Ipsen Foundation and Mathias Jucker from the University of Tübingen convened a meeting in Paris to discuss a remarkable convergence of experimental findings that I like to call the third Judgment of Paris.[1] Over the preceding four decades, there had been many meetings in Paris focused on the nature of the infectious agents causing CJD, scrapie, and later mad cow disease, but this conference sponsored by the Ipsen Foundation was to prove unique. Thirty years after the introduction of the prion concept, a dozen scientists gathered in Paris to present their findings on proteins causing a wide variety of neurodegenerative diseases.

The neurodegenerative diseases share three overarching features, distinguishing them from most other illnesses: First, more than 80 percent of the cases are sporadic, and the others are inherited. The term "sporadic" refers to the unpredictable incidence of these cases—that is, the lack of correlation with any particular risk factor except age. Besides CJD, most cases of Alzheimer's, Parkinson's, the frontotemporal dementias (FTDs), and amyotrophic lateral sclerosis (ALS, often called Lou Gehrig's disease) are sporadic, but Huntington's disease is not—it is always inherited. Second, the neurodegenerative diseases begin insidiously and progress slowly over months, years, or even decades; thus they are age dependent. Age is the greatest risk factor for such illnesses; most are rarely manifested before age forty and are referred to as late-onset maladies. Occasionally, an illness will cease to progress, as in the case of Stephen Hawking's ALS; such anomalies are labeled *forme fruste* (incomplete). Third, a particular neurodegenerative disease begins in a specific region of the nervous system and follows a distinctive pattern of spread to other areas, which is usually responsible for the progression of the illness.

Late-Onset Heritable Neurodegeneration

Cases of inherited neurodegenerative diseases give special insight into the pathogenesis of all these illnesses. Almost all familial neurodegenerative diseases are autosomal dominant disorders: That is, the DNA mutation from one parent is sufficient to cause disease in the offspring, and the mutant gene can be located on any chromosome except the X and Y sex chromosomes. Although the mutant proteins are expressed early in the development of the embryo, most neurodegenerative diseases are late-onset illnesses, including familial CJD. This finding argues that with aging some event occurs that renders a mutant disease-specific protein pathogenic. I contend that this event involves the refolding of the mutant etiologic protein into a prion state, which initiates the replication of more prions (figure 33).[2]

In contrast to the inherited cases, most patients with neurodegenerative diseases have wild-type (that is, nonmutant) prions. In these sporadic cases of prion disease, the wild-type prions form spontaneously. While a genetic mutation causes the precursor protein to be more susceptible to transformation into a prion, this same process can occur, al-

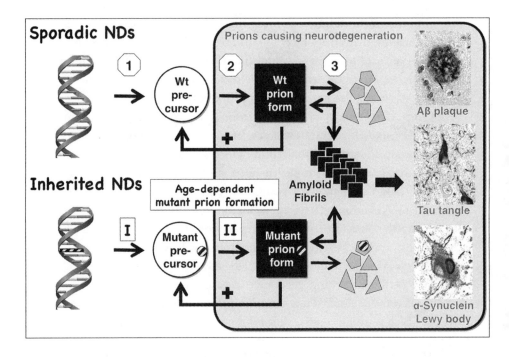

Figure 33 Neurodegeneration caused by prions. Sporadic NDs: In sporadic neurodegenerative diseases (NDs), wild-type (wt) prions multiply through self-propagating cycles of
posttranslational modification where the precursor protein (top circle) is converted into the
prion form (top square), which generally is high in β-sheet content. The small polygons
represent proteolytic cleavage products that are generated during the clearance of prions.
Depending on the protein, the fibrils coalesce into Aβ amyloid plaques in Alzheimer's
disease, neurofibrillary tangles in Alzheimer's disease or the frontotemporal dementias, or
Lewy bodies in Parkinson's disease or Lewy body dementia. Drug targets for the development of therapeutics (octagons): (1) lowering precursor protein, (2) inhibiting prion
formation, and (3) enhancing prion clearance. Inherited NDs: Late-onset heritable
neurodegeneration argues for two discrete events: the first event (arrow I) is the synthesis
of mutant precursor protein (bottom circle), and the second event is the age-dependent
formation (arrow II) of mutant prions (bottom square). The bar with diagonal lines in the
DNA helix represents mutation of a base pair within an exon, and the small circles with
diagonal lines signify the corresponding amino acid substitution.

beit at a much lower frequency, with nonmutant precursor proteins. For
example, everyone harboring a mutant PrP gene will develop familial CJD
if he or she lives long enough, but only one in a million with wild-type PrP
will develop sporadic CJD. Importantly, many of the mutant proteins

causing late-onset heritable neurodegenerative diseases are the same proteins as those found in disease-specific amyloid deposits such as plaques, neurofibrillary tangles, and Lewy bodies.[3]

To investigate the prospect that each of the proteins causing neurodegeneration can become a prion, cultured cells and Tg (transgenic) mice have been used to determine whether aggregates of such proteins can self-propagate. In each case, conditions have been identified that permit self-propagation of aggregated prion isoforms. The results of such studies are stunning: They support the contention that a prion etiology explains the late onset of both the familial and the sporadic forms of neurodegeneration.[4] Although other explanations for the late onset of the inherited forms of these disorders have been proffered, the evidence for most is meager.[5] Notably, inheritance of the ε4 allele of apolipoprotein E is the only well-established genetic risk factor for sporadic Alzheimer's disease.[6]

Huntington's disease and other afflictions with similar mutations represent a somewhat more complicated set of neurodegenerative disorders. In these illnesses, mutant genes carry an expansion of the three repeated DNA bases CAG: Each triplet encodes the amino acid glutamine. The size of the CAG expansion in the huntingtin gene has been shown to correlate inversely with the age of onset for Huntington's disease. In other words, the longer your CAG expansion, the younger you will be when you get the disease. Like the other DNA mutations causing inherited neurodegeneration, the CAG expansions are germline mutations—that is, present in the DNA of the egg or sperm. The normal huntingtin gene has thirty-six or fewer CAG repeats; the addition of four CAG repeats for a total of forty causes Huntington's disease to appear around age sixty. But adding eight more CAGs for a total of forty-eight reduces the time of disease onset by about three decades.[7]

But we still must account for the lag time between conception of the embryo and the appearance of symptoms, whether it's at age thirty or sixty. Since the mutant huntingtin protein is being expressed from an early age, some other event must be responsible for triggering the neurological dysfunction that eventually ensues. Ron Kopito of Stanford University summarized his investigations of expanded polyglutamine repeats in a fragment of the huntingtin protein. He found spontaneous formation of aggregates that self-propagated in cultured cells; in other words, they were prions.[8] These findings certainly suggest that the mutant huntingtin protein eventually becomes a prion and that its accumulation is responsible for the nervous system malfunction in Huntington's disease.

De Novo Formation of Prions

I argue that precursor proteins turn into disease-causing prions through a random process, which most of the time is a dead-end pathway where small numbers of prions are cleared via protein quality control pathways. Only when a sufficient quantity of prions is formed does the number reach a threshold level that renders the process self-propagating. Under these conditions, prion propagation becomes unregulated, and eventually, nervous system dysfunction ensues. Alternatively, a somatic mutation—that is, one that occurs in a nongermline cell—may be responsible for the spontaneous generation of prions. Under such circumstances, mutations in prion precursor proteins facilitate folding into prions, which in turn initiates self-propagation. Once enough prions accumulate to become self-perpetuating, then they begin to spread from one cell to another. It is not unreasonable to imagine that most newly formed prions never achieve sufficient number to initiate sustained replication and are eventually degraded.

In an organ such as the brain, which is composed of billions of cells, there must be a force that drives pathogenesis—the spreading from one cell to another and from region to region—of prion disease. It seems unlikely that an age-dependent slowdown or outright failure of the protein-degradation pathways is the seminal event initiating and sustaining neurodegeneration, although this has long been a favorite hypothesis to explain the pathogenesis of Alzheimer's disease. Instead, I contend that even if age-dependent inhibition of protein degradation is a contributing factor, an active process initiates and sustains the spread of prions from one neuron to another. In the 1990s, Heiko Braak, now at the University of Ulm, began reporting his studies on the spread of prions in the nervous systems of patients who died of Alzheimer's or Parkinson's. He described the spreading of Aβ amyloid plaques throughout the brains of Alzheimer's patients along neuronal tracks and also observed the migration of the tau protein in the form of neurofibrillary tangles (figure 34).[9] Recent studies have traced the spread of tau prions using magnetic resonance imaging, which can measure alterations in the circuitry of the

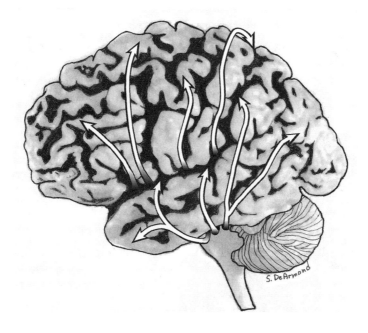

Figure 34 Spread of Aβ plaques and neurofibrillary tangles in the brains of patients with Alzheimer's disease.

brain.[10] Braak also reported the spread of Lewy bodies containing the α-synuclein prion beginning in the gut and spreading backwards up into the brain. Long before Braak's seminal studies, Bill Hadlow and others described the migration of the scrapie agent along neuronal pathways in sheep and goats, beginning in the gut and traveling up to the brain. Studies with Tg mice have likewise demonstrated the spreading of tau aggregates along neural pathways.[11]

Fungal prions provide an important perspective from which to better understand Aβ, tau, and α-synuclein prions. Reed Wickner, a thoughtful, intense, and prudent man working at the NIH, discovered fungal prions through his studies in yeast; his insights have been invaluable in defining the widening spectrum of mammalian prions.[12] The yeast prions are not infectious in the sense of being released into the culture medium and entering other yeast cells, but they are transmissible from mother to daughter cells and thus readily multiply. A segment of most yeast prion proteins is rich in the amino acids glutamine and asparagine, and this motif has proved useful in identifying some new mammalian prion proteins. More important, yeast prions provide a model for the cell-to-cell spread of certain mammalian prions including the Aβ, tau, and α-synuclein prions, which are likely to remain inside of neurons and move across synapses when spreading from cell to cell.

The early enthusiasm for transmission of two familial Alzheimer's cases to monkeys by Gajdusek and Gibbs was dashed when the inoculation results could not be repeated; in retrospect, their initial positive samples probably resulted from CJD cases being mislabeled Alzheimer's. Their suspicions were correct, but the tools to test their hypothesis were inadequate. Building on George Glenner's discovery of the Aβ peptide, Harry Baker and Rosalind Ridley, two of my collaborators in the inherited human PrP prion disease story (see chapter 12), working at Cambridge University, inoculated marmosets with brain homogenates from deceased Alzheimer's disease patients.[13] After more than three and a half years, the marmosets developed Aβ amyloid plaques and perivascular Aβ deposits. The work of Baker and Ridley showed that Aβ prions in Alzheimer's brain homogenates multiplied in marmoset brains and spread from one region to another, presumably along neuronal tracks. As in humans, increasing levels of prions in the one region of the marmoset brain appears to be the force that induces movement to other regions. The spread of Aβ prions became easier to trace when rodent models became available.

At the Paris meeting, Mathias Jucker and Lary Walker summarized their Tg mouse model for measuring Aβ prions (figure 35). They reported the transmission of Aβ prions from the brains of Tg mice that accumulated large numbers of Aβ amyloid plaques, as well as from the brains of humans who died of Alzheimer's disease. While the initial studies were performed by injections of crude homogenates containing Aβ prions into the brains of Tg mice, more work showed that inoculations into the bellies of mice also works.[14]

With Erdem Tamgüney, an intense and thoughtful postdoctoral fellow, I first asked whether Tg mice that can emit light like fireflies might facilitate studies of scrapie prions. We began by measuring the light coming through the skulls of these mice during prion replication in the brain. The technique worked well and allowed us to detect replication of scrapie prions a couple of months before signs of neurological

Figure 35   Lary Walker (left) and Mathias Jucker study prions causing Alzheimer's disease. Alois Alzheimer's picture is displayed in the background, on a poster celebrating the one hundredth anniversary in 2006 of the first description of this disease.

Figure 36   Working with me (left to right), Jan Stöhr, Kurt Giles, and Joel Watts study prions causing CJD, Alzheimer's disease, Parkinson's disease, and the frontotemporal dementias.

dysfunction appeared.[15] On the transmissibility of Alzheimer's, I worked closely with Kurt Giles, a talented and careful UCSF faculty member, as well as two gifted and imaginative postdoctoral fellows, Jan Stöhr and Joel Watts (figure 36). We asked whether measuring light emitted from the brain could also act as a reporter in Tg mouse models of Alzheimer's.[16]

Interlude: Bioluminescence Imaging of Transgenic Mice

The biochemical emission of light called bioluminescence occurs in fireflies when the enzyme luciferase catalyzes the breakdown of the chemical compound luciferin. The mouse skull is sufficiently thin that light coming from the brain can be easily measured in live animals. To adapt this to studies of prions in mice, we obtained Tg mice expressing luciferase under control of the gene regulator for the glial fibrillary acidic protein—or GFAP, as it is generally called. The GFAP gene is activated in response to many different brain insults. To determine whether bioluminescence imaging could be used to monitor the progression of scrapie, we injected PrP^{Sc} prions into Tg mice and saw a sustained increase in bioluminescence from mouse brains after two months, instead of having to wait four months for signs of neurological dysfunction to appear.*

I decided to determine whether bioluminescence imaging could be used to monitor prions causing other neurodegenerative diseases including Alzheimer's, Parkinson's, and the FTDs. We began by breeding the mice expressing luciferase with models of inherited Alzheimer's, Parkinson's, and FTD. With luciferase, we could monitor the development of neurodegeneration in living animals. The key was to pick Tg mice expressing a mutant, disease-specific protein that caused spontaneous neurodegeneration in older mice. The timing of these experiments had to be just right so that we could inoculate young mice and demonstrate that an acceleration of disease was due to the injected prions. Once bigenic mice expressing both luciferase and a mutant, disease-specific protein became available, we inoculated them with homogenates prepared from the brains of patients who died of these disorders, or from older Tg mice showing spontaneous neurodegeneration.

We measured incubation times by determining the interval from inoculation to a sustained increase in light emitted from the brain. Inoculation of brain

---

*G. Tamgüney, K. P. Francis, K. Giles, A. Lemus, S. J. DeArmond, and S. B. Prusiner, "Measuring prions by bioluminescence imaging," *Proc. Natl. Acad. Sci. USA* 106: 15002–6 (2009).

homogenates from Alzheimer's patients and Tg mice, as well as purified Aβ and synthetic Aβ amyloid fibrils, all shortened the incubation times as measured by bioluminescence.[*] Brain homogenates from old Tg mice expressing mutant human α-synuclein produced an increase in luminescence when inoculated into bigenic mice expressing luciferase and mutant human α-synuclein.[†] Similar results were obtained when brain homogenates were prepared from patients who died of multiple system atrophy and were injected into the brains of the bigenic mice.

[*]J. C. Watts, K. Giles, S. K. Grillo, A. Lemus, S. J. DeArmond, and S. B. Prusiner, "Bioluminescence imaging of Abeta deposition in bigenic mouse models of Alzheimer's disease," *Proc. Natl. Acad. Sci. USA* 108: 2528–33 (2011); J. Stöhr, J. C. Watts, Z. L. Mensinger, A. Oehler, S. K. Grillo, S. J. DeArmond, S. B. Prusiner, and K. Giles, "Purified and synthetic Alzheimer's amyloid beta (Aβ) prions," *Proc. Natl. Acad. Sci. U.S.A.* 109: 11025–30 (2012).

[†]J. C. Watts, K. Giles, A. Oehler, L. T. Middleton, D. T. Dexter, S. M. Gentleman, S. J. DeArmond, and S. B. Prusiner, "Transmission of multiple system atrophy prions to transgenic mice," *Proc. Natl. Acad. Sci. U.S.A.* 110: 19555–60 (2013).

Following Jucker and Walker's lead, we performed a series of genetic crosses in order to build double Tg mice—bigenic mice that could both replicate Aβ prions and emit a light signal when the prions were multiplying. We were able to detect changes in the brains of these mice inoculated with samples highly enriched for Aβ peptide. But as in our earlier studies with scrapie prions,[17] we wanted to be sure that Aβ, and not something else in the sample, was responsible for the light signal that we were measuring. To do this, we obtained Aβ peptides that had been synthesized from individual amino acids in the absence of any cells or larger organisms. Could the synthetic Aβ peptide alone cause disease in our bigenic mice? I presented the results of these experiments at the Paris meeting: The Aβ peptides had indeed transmitted neurodegeneration to our bigenic mice.[18]

The breeding of the bigenic mice for these studies was complicated and had taken several years of work. While these studies were in progress, investigators from other laboratories described evidence for prions in Alzheimer's, Parkinson's, and frontotemporal dementias using either mice or cultured cells. Although their data mounted convincing arguments for prions as the cause of each of the human neurodegenerative diseases, I still needed my own data to be persuaded. I feared I might be too eager to accept the conclusions of others, since their findings provided evidence for conjectures I had made almost three decades earlier (see chapter 9). I also worried that some systematic artifact had led them to an erroneous interpretation of their experiments. Our light-emitting technique, called bioluminescence imaging, provided a novel, independent approach and reassured me that prions do indeed feature in the pathogenesis of many different neurodegenerative diseases.

At the Paris meeting, Michel Goedert of Cambridge University described the transmission of mutant tau prions produced in Tg mice to recipient mice expressing wild-type human tau.[19] After about a year, the inoculated mice showed tau aggregates in their brains. Goedert's findings complement those of Marc Diamond of Washington University in St. Louis, who found that a segment of the tau protein could be aggregated and introduced into cells, where it initiated tau prion formation—studies he performed while still an assistant professor of neurology at UCSF.[20]

Interlude: Synthetic PrP Prions

My group had been working on the production of synthetic PrP prions since the early 1990s. After all, this was the most convincing way to show that PrP prions consisted only of PrP^{Sc} molecules and that no other components were buried inside the prion, such as a small stretch of DNA or RNA. The first glimmer of a synthetic PrP prion came in studies performed by Kiyotoshi Kaneko, a thoughtful, unassuming postdoctoral fellow, who used a synthetic PrP peptide of fifty-five amino acids containing the GSS mutation P102L—a peptide we produced by coupling individual amino acids.* After Kaneko mixed the PrP peptide with acetonitrile, which caused it to adopt β-rich structure, he injected it into a Tg mouse model of GSS that Karen Hsiao had created (see chapter 12). About a year later, the inoculated mice became ill; the uninoculated mice remained well for another six months. Detractors argued that subsequent illness in the uninoculated controls meant that the peptide was not a prion but merely an accelerator of disease. Such arguments persisted for a more than a decade.† But we showed, by serial passage, that the brains of the inoculated mice contained a transmissible "entity"—that is, a prion.‡

*K. Kaneko, H. L. Ball, H. Wille, H. Zhang, D. Groth, M. Torchia, P. Tremblay, J. Safar, S. B. Prusiner, S. J. DeArmond, M. A. Baldwin, and F. E. Cohen, "A synthetic peptide initiates Gerstmann-Sträussler-Scheinker (GSS) disease in transgenic mice," *J. Mol. Biol.* 295: 997–1007 (2000).

†K. E. Nazor, F. Kuhn, T. Seward, M. Green, D. Zwald, M. Purro, J. Schmid, K. Biffiger, A. M. Power, B. Oesch, A. J. Raeber, and G. C. Telling, "Immunodetection of disease-associated mutant PrP, which accelerates disease in GSS transgenic mice," *EMBO J.* 24: 2472–80 (2005).

‡P. Tremblay, H. L. Ball, K. Kaneko, D. Groth, R. S. Hegde, F. E. Cohen, S. J. DeArmond, S. B. Prusiner, and J. G. Safar, "Mutant PrP^{Sc} conformers induced by a synthetic peptide and several prion strains," *J. Virol.* 78: 2088–99 (2004).

Following in part the lead of Jonathan Weissman, a gifted professor at UCSF who works on yeast prions, we took PrP made in bacteria and assembled it into amyloid fibrils. Ilia Baskakov, an energetic postdoctoral fellow working with Fred Cohen and me, began his experiments by producing the fibrils and injecting them into mice. When the mice had remained well for more than 250 days, we concluded that the experiment was negative and wrote up our findings.[*] While I believed that the experiment had been terminated, Giuseppe Legname, an enthusiastic assistant professor working with me, brought the continued studies to my attention a year later. About 600 days after inoculation, the mice developed signs of neurological dysfunction, and their brains showed the hallmarks of scrapie.[†] When Legname examined the stability of a dozen prion strains in mice, including an isolate of one synthetic prion, we found that the more stable strains, which resisted denaturation, produced longer incubation times than the less stable ones.[‡]

Based on this finding, David Colby, a gifted postdoctoral fellow, produced synthetic prions under conditions creating PrP preparations highly resistant to denaturation, along with some that were highly susceptible and some that were intermediate. The stable preparations produced long incubation times, whereas the more labile ones yielded much shorter incubation times.[§] Although these investigations extended our earlier work, they were exceedingly expensive because of the long incubation times. More recent studies, using Tg mice expressing a mutant form of PrP lacking a glycolipid anchor, have shown that these animals are superior for detecting synthetic prions.[¶] Another approach to shortening the incubation time involves the use of Tg mice expressing bank vole PrP.[**]

[*]I. V. Baskakov, G. Legname, M. A. Baldwin, S. B. Prusiner, and F. E. Cohen, "Pathway complexity of prion protein assembly into amyloid," *J. Biol. Chem.* 277: 21140–48 (2002).

[†]G. Legname, I. V. Baskakov, H.-O. B. Nguyen, D. Riesner, F. E. Cohen, S. J. DeArmond, and S. B. Prusiner, "Synthetic mammalian prions," *Science* 305: 673–76 (2004); G. Legname, H.-O. B. Nguyen, I. V. Baskakov, F. E. Cohen, S. J. DeArmond, and S. B. Prusiner, "Strain-specified characteristics of mouse synthetic prions," *Proc. Natl. Acad. Sci. U.S.A.* 102: 2168–73 (2005).

[‡]G. Legname, H.-O. B. Nguyen, D. Peretz, F. E. Cohen, S. J. DeArmond, and S. B. Prusiner, "Continuum of prion protein structures enciphers a multitude of prion isolate-specified phenotypes," *Proc. Natl. Acad. Sci. USA* 103: 19105–10 (2006).

[§]D. W. Colby, K. Giles, G. Legname, H. Wille, I. V. Baskakov, S. J. DeArmond, and S. B. Prusiner, "Design and construction of diverse mammalian prion strains," *Proc. Natl. Acad. Sci. USA* 106: 20417–22 (2009).

[¶]J. Stöhr, J. C. Watts, G. Legname, A. Oehler, A. Lemus, H.-O. B. Nguyen, J. Sussman, H. Wille, S. J. DeArmond, S. B. Prusiner, and K. Giles, "Spontaneous generation of anchorless prions in transgenic mice," *Proc. Natl. Acad. Sci. USA* 108: 21223–28 (2011).

[**]J. C. Watts, K. Giles, J. Stöhr, A. Oehler, S. Bhardwaj, S. K. Grillo, S. Patel, S. J. DeArmond, and S. B. Prusiner, "Spontaneous generation of rapidly transmissible prions in transgenic mice expressing wild-type bank vole prion protein," *Proc. Natl. Acad. Sci. U.S.A.* 109: 3498–503 (2012).

Many of the principles learned from the creation of synthetic PrP prions are proving applicable to work on prions causing Alzheimer's, Parkinson's, and the frontotemporal dementias.*

*J. Stöhr, J. C. Watts, Z. L. Mensinger, A. Oehler, S. K. Grillo, S. J. DeArmond, S. B. Prusiner, and K. Giles, "Purified and synthetic Alzheimer's amyloid beta (Aβ) prions," *Proc. Natl. Acad. Sci. U.S.A.* 109: 11025–30 (2012).

New Interface Between Psychiatry and Neurology

Many of the frontotemporal dementias at the interface between psychiatry and neurology are called tauopathies. Most FTD patients have a sporadic form of the illness; when their tau genes were sequenced, no mutations were found. However, FTD patients in families with a history of FTDs do often harbor tau gene mutations.[21] Notably, postmortem examinations of the brains of such patients do not exhibit Aβ amyloid plaques, demonstrating that tau gene mutations do not cause Alzheimer's disease.

Psychiatrists see FTD patients for several years before referring them to neurologists with the diagnosis of Alzheimer's disease. Tau lesions in the frontal lobes can produce disinhibition characterized by behavioral changes that include apathy, inappropriate social interactions, depression, insomnia, and diminished executive functions.[22] Later symptoms include difficulty finding words, drug abuse, alcoholism, and sometimes suicide.

Numerous neurofibrillary tangles filled with tau have been found in the frontal lobes of contact sport athletes, some of whom have committed suicide. Although the first football player in whom a tauopathy was recognized played professional football for sixteen seasons and died at age fifty,[23] high school and college players with tauopathies have been identified.[24] In 2011, a twenty-seven-year-old Marine who suffered multiple episodes of traumatic brain injury during explosions of roadside bombs in Iraq was found to suffer from a tauopathy. Following the diagnosis of posttraumatic stress disorder, he was honorably discharged. Subsequently, he divorced his wife and became an alcoholic. Eight months after discharge, he hanged himself, and an autopsy found numerous neurofibrillary tangles in his frontal lobes.[25] The frontal lobes of four more soldiers with posttraumatic stress disorder who had been exposed to improvised explosive devices showed focal collections of neurofibrillary tangles on postmortem examination.[26] One shot himself, another died after an overdose of prescription analgesics, the third died of a ruptured aneurysm, and the fourth of an intracerebral hemorrhage.

Studies of posttraumatic FTD are challenging, because clinical symptoms can appear decades after the subject experienced a traumatic brain injury. The number of soldiers suffering from posttraumatic stress disorder who have FTD is unknown—as is the number of episodes of traumatic brain injury needed to induce FTD.[27] It seems likely that the number of episodes will vary from one person to another and will also depend on the type and extent of the brain injury. The U.S. government reports one suicide per day in active-duty personnel and twenty-two in veterans.[28] At present, we do not know how many of these have tau prions in their frontal lobes. It is crucial to elucidate the mechanism by which trauma provokes the conversion of normal tau into prions.

While the foregoing illnesses are generally called primary tauopathies, the most common is the secondary tauopathy of Alzheimer's disease. It seems likely that the initial event in Alzheimer's disease is the formation of Aβ prions in sufficient number to initiate self-propagation. Considerable experimental evidence suggests that Aβ prions stimulate the transformation of normal tau into a prion.[29] Other examples of secondary tauopathies include rabies, postencephalitic Parkinson's, and Niemann-Pick C disease. As in familial Alzheimer's, seven different mutant human CJD prions induce large numbers of neurofibrillary tangles.[30]

Understanding the structural transitions in tau that occur in sporadic and inherited cases of FTD will be critical in developing effective drugs and informative molecular diagnostics. We know that tau becomes rich in β-sheet as it folds into a prion and that the tau prions assemble into long fibrils that coalesce into tangles, but we lack the detailed structural information needed for effective drug design. We also need to learn how head trauma, Aβ, and PrP^{Sc} prions (as well as complex lipids and rabies viruses) provoke tau proteins to refold into prions.

Unexpectedly, more and more causes of FTDs are being discovered. Besides tau, at least six other proteins have been found to cause FTDs. Among these, the TDP43 and FUS proteins are particularly notable, and both seem likely to form prions.

Interlude: Expanding Spectrum of FTDs

FTD phenotypes include progressive supranuclear palsy, corticobasal degeneration, Pick's disease, and argyrophilic grain disease, each of which appears to be caused by a different strain of tau prions producing lesions in different neuronal circuits.[*] Progressive supranuclear palsy was initially called Steele-Richardson-Olszewski syndrome, reflecting the first clinical-pathological description of this disorder.[†]

While the conversion of tau into a prion is a frequent cause of FTD, at least six other proteins also cause FTDs—disorders that sometimes exhibit signs of Parkinson's disease.[‡] These proteins are C9orf72, TDP43, FUS, progranulin, vasolin-containing protein, and charged multivesicular body protein 2B. Some patients with FTD also develop ALS.

By combining genetic studies with neuropathologic investigations, medical scientists have been able to tease out the causes of these once mysterious illnesses. Studies of families with FTD have pointed to a single gene whose mutation in DNA results in an aberrant protein. Once the protein is identified, we can use antibodies to examine brains from deceased members of that and other FTD families for the presence of deposits composed of that particular protein. This approach has revealed accumulations of TDP43 and FUS proteins in the brains of FTD family members with mutations in the genes encoding either of these proteins; they are both RNA-binding proteins and rich in glutamine and asparagine residues like those in yeast proteins that form prions.[§] The formation of TDP43 or FUS prions would create a second event that explains the late onset of these inherited FTDs. But in the case of TDP43, despite the tantalizing data suggesting that it may form prions, the presence of TDP43 deposits in a wide range of neurological disease argues that our understanding of this protein is still incomplete. Moreover, some sporadic forms of FTD, often with ALS, exhibit deposits of TDP43, but the patients have no mutations in their TDP43 gene.

Different pathogenic mechanisms feature in FTD caused by mutation in the progranulin and C9orf72 genes. Mutation in either gene is thought to re-

---

[*]W. W. Seeley, R. K. Crawford, J. Zhou, B. L. Miller, and M. D. Greicius, "Neurodegenerative diseases target large-scale human brain networks," *Neuron* 62: 42–52 (2009).

[†]J. C. Steele, J. C. Richardson, and J. Olszewski, "Progressive supranuclear palsy: A heterogeneous degeneration involving the brain stem, basal ganglia and cerebellum with vertical gaze and pseudobulbar palsy, nuchal dystonia and dementia," *Arch. Neurol.* 10: 333–59 (1964).

[‡]R. Rademakers, M. Neumann, and I. R. Mackenzie, "Advances in understanding the molecular basis of frontotemporal dementia," *Nat. Rev. Neurol.* 8: 423–34 (2012).

[§]O. D. King, A. D. Gitler, and J. Shorter, "The tip of the iceberg: RNA-binding proteins with prion-like domains in neurodegenerative disease," *Brain Res.* 1462: 61–80 (2012).

sult in a deficiency of the protein. In older patients with a progranulin deficiency, TDP43 deposits become abundant in the brain, but it is unknown whether TDP43 prions form in response to progranulin deficiency.[*] An expansion of a GGGGCC hexanucleotide repeat in the C9orf72 gene has been identified as the most common pathogenic mutation in families with autosomal dominant FTD and/or ALS.[†] Patients with a mutation in the C9orf72 gene produce an aberrant protein composed of only four amino acids that aggregates to form inclusion bodies in the cytoplasm of neurons.[‡] There is an irony to this discovery: Back when I was considering all the possibilities for infectious proteins, I speculated that prions might be proteins composed of small repeating units of only a few amino acids, as is the case for some nonribosomal polypeptide antibiotics; that idea was discarded when we found that PrP contained all twenty amino acids (see chapter 10). Whether or not this unusual protein, which is encoded upstream of the coding region of the C9orf72 gene, becomes a prion before aggregating within neurons remains to be determined.

[*]R. Rademakers, M. Baker, J. Gass, J. Adamson, E. D. Huey, P. Momeni, S. Spina, G. Coppola, A. M. Karydas, H. Stewart, N. Johnson, G. Y. Hsiung, B. Kelley, K. Kuntz, E. Steinbart, E. M. Wood, C. E. Yu, K. Josephs, E. Sorenson, K. B. Womack, S. Weintraub, S. M. Pickering-Brown, P. R. Schofield, W. S. Brooks, V. M. Van Deerlin, J. Snowden, C. M. Clark, A. Kertesz, K. Boylan, B. Ghetti, D. Neary, G. D. Schellenberg, T. G. Beach, M. Mesulam, D. Mann, J. Grafman, I. R. Mackenzie, H. Feldman, T. Bird, R. Petersen, D. Knopman, B. Boeve, D. H. Geschwind, B. Miller, Z. Wszolek, C. Lippa, E. H. Bigio, D. Dickson, N. Graff-Radford, and M. Hutton, "Phenotypic variability associated with progranulin haploinsufficiency in patients with the common 1477C—>T (Arg493X) mutation: an international initiative," Lancet Neurol 6: 857–68 (2007).

[†]E. Majounie, A. E. Renton, K. Mok, E. G. Dopper, A. Waite, S. Rollinson, A. Chio, G. Restagno, N. Nicolaou, J. Simon-Sanchez, J. C. van Swieten, Y. Abramzon, J. O. Johnson, M. Sendtner, R. Pamphlett, R. W. Orrell, S. Mead, K. C. Sidle, H. Houlden, J. D. Rohrer, K. E. Morrison, H. Pall, K. Talbot, O. Ansorge, D. G. Hernandez, S. Arepalli, M. Sabatelli, G. Mora, M. Corbo, F. Giannini, A. Calvo, E. Englund, G. Borghero, G. L. Floris, A. M. Remes, H. Laaksovirta, L. McCluskey, J. Q. Trojanowski, V. M. Van Deerlin, G. D. Schellenberg, M. A. Nalls, V. E. Drory, C. S. Lu, T. H. Yeh, H. Ishiura, Y. Takahashi, S. Tsuji, I. Le Ber, A. Brice, C. Drepper, N. Williams, J. Kirby, P. Shaw, J. Hardy, P. J. Tienari, P. Heutink, H. R. Morris, S. Pickering-Brown, and B. J. Traynor, "Frequency of the C9orf72 hexanucleotide repeat expansion in patients with amyotrophic lateral sclerosis and frontotemporal dementia: A cross-sectional study," Lancet Neurol 11: 323–30 (2012).

[‡]K. Mori, S. M. Weng, T. Arzberger, S. May, K. Rentzsch, E. Kremmer, B. Schmid, H. A. Kretzschmar, M. Cruts, C. Van Broeckhoven, C. Haass, and D. Edbauer, "The C9orf72 GGGGCC repeat is translated into aggregating dipeptide-repeat proteins in FTLD/ALS," Science 339: 1335–38 (2013).

Parkinson's Disease and Synuclein Prions

At the Ipsen Conference, Patrik Brundin, from Lund University in Sweden, reported earlier studies on Lewy bodies[31] in grafted fetal neurons, a decade after their transplantation into Parkinson's patients,[32] suggesting that a major protein of the Lewy body might have turned into a prion. Simultaneous with Brudin's publication of European patients,

Warren Olanow and Jeff Kordower reported similar findings in Parkinson's patients in the United States.[33] Lewy bodies are pathognomonic for Parkinson's disease; the surface of a Lewy body is composed of fibrils containing α-synuclein (figure 37). In earlier studies, some familial cases of Parkinson's disease were found to be due to a mutation in the α-synuclein gene, and this finding led to the discovery that Lewy bodies possess a halo of radiating fibrils composed of α-synuclein (see figure 33).[34] Taken together, the finding of Lewy bodies in fetal grafts suggested that α-synuclein prions crossed from the patients' own neurons into the grafted neurons and induced a structural change in α-synuclein.[35] Once established, this process became self-propagating, like all prions. In earlier studies, Braak described the movement of aggregated α-synuclein from the gut to the brain, suggesting that at least some α-synuclein prions might be formed in the intestine either spontaneously or from exposure to some environmental insult.[36]

Besides accumulating in Parkinson's disease, aggregated α-synuclein accumulates in Lewy body dementia and multiple system atrophy (once called Shy-Drager syndrome after my Penn neurology professor Milton Shy); Parkinson's signs and symptoms are present in both of these disorders. In Lewy body dementia, fluctuating cognition with great variations in attention and alertness is a prominent feature, as are visual hallucinations. In multiple system atrophy, signs of autonomic nervous system dysfunction are prominent and include orthostatic hypotension, resulting in dizziness or fainting upon standing up, urinary incontinence, impotence, constipation, and dysregulation of body temperature due to hypohidrosis (deficient sweating). While α-synuclein aggregates in Lewy body dementia form Lewy bodies within neurons, the α-synuclein aggregates in multiple system atrophy are found inside oligodendrocytes, which produce a form of electrical insulation called myelin sheaths that facilitate axon functioning throughout the central nervous system.

Another Paris participant, Virginia Lee from the University of Pennsylvania, described her studies of α-synuclein prions in Tg mice and cultured cells. Ten years earlier, Lee had created Tg mice expressing mutant human α-synuclein that caused neurological disease when they were about four hundred days old. When brains from these ill, uninoculated mice were homogenized and injected into weanling mice in her own laboratory or that of Thierry Baron, working in Lyon,[37] the recipient mice became ill at about two hundred days of age. This accelerated neurodegeneration in the inoculated Tg mice demonstrated that there were α-synuclein prions in the inocula prepared from the brains of uninoculated Tg mice. Lee also described her studies showing aggregation of α-synuclein in cultured cells.[38] Subsequently, Lee and her colleagues reported the transmission of synthetic α-synuclein prions into wild-type mice, as did Masato Hasegawa at the Tokyo Metropolitan Institute of Medical Science.[39]

Unexpectedly, Joel Watts, Kurt Giles, Lefkos Middleton of Imperial College, and I found that homogenates prepared from the brains of patients who died of multiple system atrophy caused progressive neurological dysfunction in Tg mice expressing mutant human α-synuclein in fewer than one hundred days after inoculation.[40] We concluded that brains from patients who died of multiple system atrophy contained aggressive strains of α-synuclein prions. In fact, these prions represent the first new human prions that kill laboratory animals to be discovered in almost a half-century. In the late 1960s, the experimental transmission of kuru and CJD to apes and monkeys was the first report of human prions that killed an experimental host (see chapter 5).

More than half of people with Parkinson's disease exhibit dementia toward the end of their lives. At autopsy, the brains of these demented people with Parkinson's contain numerous Lewy bodies and often Aβ amyloid plaques and tangles. Teasing out the relative contributions of α-synuclein, Aβ, and tau to dementia-Parkinson's patients are likely to require major advances in neuroimaging. Using bigenic mice and rats, it may be possible to determine whether α-synuclein prions induce the formation of Aβ and possibly tau prions or vice versa.

Other presentations in Paris included investigations of prions causing ALS. Anne Bertolotti from Cambridge University reported her studies on superoxide dismutase 1, an enzyme present in virtually all cells that functions as an antioxidant. She and others found that aggregates of mutant human superoxide dismutase 1 are self-propagating and as such are prions.[41] More than sixty different mutations in superoxide dismutase 1 have been identified as a cause of familial ALS.

The Paris meeting ended with an exciting talk by Eric Kandel of Columbia University, who described his studies of a prion that seems to control localized gene expression in long-term memory.[42] In addition to this memory prion, two other mammalian proteins have been found to form prions but do not appear to cause disease. Like the memory prion, these prions seem to perform important cellular functions, and all three contain glutamine-rich domains like those of yeast prions.[43]

A Remarkable Convergence

The convergence of studies demonstrating that prions feature in the pathogenesis of common neurodegenerative maladies has created a profound change in thinking about these devastating illnesses.[44] Many mysteries are now explicable within the framework of the prion concept. Generally, neurotoxic prions are oligomers (or small clumps of a few prions), not large collections of amyloid fibrils such as plaques, tangles, or Lewy bodies (figure 37). For example, PrP amyloid plaques are a non-obligatory feature of CJD;[45] moreover, Aβ plaques do not correlate with the severity of dementia while the levels of oligomeric Aβ prions are related to the extent of the cognitive deficits.[46] It is not surprising that elevated bioluminescence and behavioral changes precede the appearance of Aβ plaques in Tg mice.[47] Studies with cultured neurons showed the transit of Aβ oligomers from one cell to another.[48]

Most important, strategies for developing informative molecular diagnostics and effective therapeutics for these elusive disorders seem likely to emerge from our knowledge of prions. Early diagnosis will require identification of prions long before symptoms appear. Meaningful treatments will probably require cocktails of drugs to diminish the precursor protein, interfere with its conversion into prions, and enhance their clearance.[49]

On a more personal note, I must admit to a certain satisfaction in watching a discovery made in the early 1980s[50] play out over three decades. Our finding that PrP^{Sc} prions assemble into amyloid fibrils and display the ultrastructural features and tinctorial properties of amyloid argued that many other neurodegenerative disorders besides scrapie and CJD were likely to be caused by prions. But it took the talent, imagination, and hard work of hundreds of scientists to create the tools that led to the current understanding of neurodegeneration.[51]

The Greek poet Archilochus wrote, "The fox knows many things, but the hedgehog knows one big thing." This dichotomy was expanded upon in one of Isaiah Berlin's most famous essays, in which he points out that hedgehogs see the world through the lens of a single idea, whereas foxes

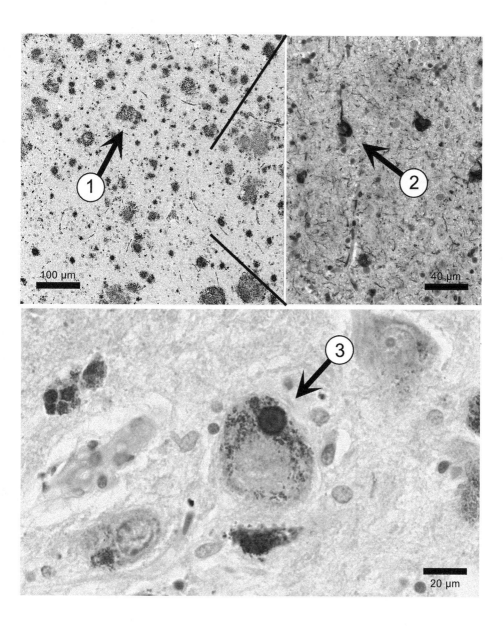

Figure 37 Prion deposits in the brains of people who died of Alzheimer's and Parkinson's diseases. In Alzheimer's, Aβ peptides formed amyloid fibrils that coalesced into plaques (1). These Aβ prions also stimulated tau to become prions, which polymerize into fibrils that condensed to form threads and neurofibrillary tangles (2). The lines indicate the approximate area from which the right micrograph was taken. In Parkinson's disease, the synuclein protein changes its shape and after becoming a prion, it polymerizes into fibrils; the synuclein fibrils decorate the surface of Lewy bodies (3).

act on a wide range of experiences and do not reduce the world to one context.[52] A "hedgehog" view of neurodegeneration has long been unpopular, but the foxes now seem to be losing ground. Rarely does a single idea triumph in science; seldom do we get a glimpse of a unifying concept. It took more than a quarter of a century for the prion concept to emerge from the shadows cast by so many confounding observations. But we are now at the point where the burden of proof is beginning to shift from those who claim prions cause such common illnesses as Alzheimer's and Parkinson's to those who deny this etiology.

Epilogue:
The Quest for Therapeutics

Y ou have to give people some hope. Tell them that they can do something to prevent Alzheimer's. Why can't they take vitamins and supplements or do brain exercises that will help prevent dementia?" I had just finished speaking at the Aspen Institute; my host and friend, Kenneth Adelman, was admonishing me about being too set in my ways. "Stan, you need to give them some reason to believe that Alzheimer's is not hopeless—otherwise, they won't be interested in supporting your research."

I told Ken that I believed biomedical science would eventually come up with effective medications but that it is misleading to tell people that modifying their diet, doing crossword puzzles, or learning a foreign language would protect them from Alzheimer's when there was no meaningful evidence to support such an assertion. I then explained to him that distorting my understanding to fit other people's desires was dangerous. If you say something often enough, you might begin to believe it. For a scientist, the most important trait is intellectual honesty within himself.

In a world where truth tends to be increasingly perverted, scientists are experiencing increasing difficulty teaching their students that the most important truth lies within themselves. Too often, scientists are tempted to interpret their data as supporting their favorite hypothesis and are reluctant to accept a seemingly disappointing conclusion.[1]

Ken was also uneasy about my telling people that there is not a single medicine that halts or even slows any neurodegenerative disease. But I had to be truthful. While Aricept and its copies generate revenues of about $3 billion annually, these drugs do not retard the relentless progression of Alzheimer's disease. However, I did insist during my talk that I firmly believed that scientists would someday discover drugs for effective treatment and even the prevention of Alzheimer's—and that such research was both plausible and reasonable. Modern drug discovery is a well-developed science and deserves substantial support.

In 1994, at age eighty-three and five years after leaving the presidency, Ronald Reagan was diagnosed with Alzheimer's; how much earlier his memory deficits became manifest is unclear. Reagan's mother died of Alzheimer's at age seventy-nine. Shortly after learning about his illness, Reagan wrote a courageous letter to the American people:

> I have recently been told that I am one of the millions of Americans who will be afflicted with Alzheimer's disease. . . .
>
> At the moment I feel just fine. I intend to live the remainder of the years God gives me on this earth doing the things I have always done. I will continue to share life's journey with my beloved Nancy and my family. I plan to enjoy the great outdoors and stay in touch with my friends and supporters.
>
> Unfortunately, as Alzheimer's disease progresses, the family often bears a heavy burden. I only wish there was some way I could spare Nancy from this painful experience. When the time comes, I am confident that with your help she will face it with faith and courage.
>
> In closing let me thank you, the American people, for giving me the great honor of allowing me to serve as your president. . . .
>
> I now begin the journey that will lead me into the sunset of my life. I know that for America there will always be a bright dawn ahead.

Ronald Reagan and his family suffered for ten years as his dementia progressively worsened. In 2004, Reagan died of Alzheimer's at the age of ninety-three. It is disappointing that Reagan's name is unlikely to be associated with a cure for Alzheimer's. His family and friends could have demanded a "War on Alzheimer's" had they chosen to do so. I remain disheartened that Nancy Reagan failed to mount an assault on Alzheimer's disease, especially since she came from such a neurologically sophisticated family. Her father Loyal Davis and brother Richard were both professors of neurosurgery: Loyal at Northwestern University, where he was chairman of the Department of Surgery for thirty-one years, and Richard at the University of Pennsylvania (he was already at Penn when I was a medical student). To be fair, Nancy Reagan was an advocate for stem-cell research, which she believed would lead to a cure for Alzheimer's.

Imagine a foreign country assaulting our nation and killing 500,000 Americans. We would commit hundreds of billions of dollars to vanquish the enemy. Yet 500,000 Americans die of Alzheimer's annually, and we continue to allocate only 1.5 percent of the budget of the National Institutes of Health to Alzheimer's research. In contrast, cancer research consumes more than 22 percent of that budget.

Straightening Out the Numbers on Alzheimer's Disease

The economics of Alzheimer's do not make sense. We spend nearly $200 billion annually caring for Alzheimer's victims, a colossal sum that includes the lost productivity both of patients and caregivers.[2] Moreover, Alzheimer's victims occupy half of all nursing home beds in America. Yet we devote less than half a billion dollars a year to research, so we spend four hundred times more on care than on research directed at stopping this curse. This makes no sense.

The statistics of Alzheimer's are no less irrational than its economics. Currently, 5.3 million Americans suffer from Alzheimer's, and 500,000 new cases appear each year. Alzheimer's oscillates with diabetes as the sixth- or seventh-leading cause of death in the United States, according to death certificate statistics from the Centers for Disease Control (CDC), with about 70,000 Alzheimer's victims dying annually. But such numbers make no sense when there are half a million new cases annually. The answer lies in the underreporting of dementia on death certificates. Several studies reporting on the clinical diagnosis of dementia in living patients have concluded that Alzheimer's was underrecorded by about 75 percent on death certificates.[3] I confirmed this discrepancy when I asked the CDC to tally all deaths that included a dementing disease. Among the twenty disorders, the result was a mere 104,000 deaths.

It seemed to me that if only a little more than 100,000 people died with Alzheimer's annually in America, then the number of living, demented people should be rising rapidly. When I convened a group of neuroepidemiologists including David Bennett from Rush Medical College, Kenneth Langa from the University of Michigan, and Kristine Yaffe from UCSF, they found that about half a million people die annually with Alzheimer's and that the disease shortened life spans by several years compared with those of people who were cognitively normal.[4] These findings argue that Alzheimer's is one of the top three killer diseases in America—the other two being heart disease and cancer.

With Sandra Day O'Connor (whose husband died of Alzheimer's) and the gerontologist Ken Dychtwald, I coauthored an October 2010 op-ed piece in the *New York Times* on the rapidly growing number of people with Alzheimer's disease.[5] In response, Kathleen Hart from Longview, Texas, wrote:

> As a woman with Alzheimer's disease, I read the article with interest and appreciation. Further awareness needs to be raised about those of us who receive a diagnosis of Alzheimer's at a young age—I'm 53—and the plight of those of us who, for one reason or another, have no one to act as a caregiver, when the time comes. I think that it is easy to forget (no pun intended) about those of us with Alzheimer's, because our suffering is "invisible." None of my neighbors see me going to the hospital for any type of treatment, for example, not because there is nothing wrong with me, as they think, but because I have a disease for which no treatment is available. No one at the grocery store I go to sees me in a wheelchair, or wearing a sling, which leads them to think I'm healthy, while, because of the deterioration of my brain, I'm having a hard time making a decision about what type of cereal to buy. In the later stages of the disease, I'll also be "invisible," because I'll be confined to a nursing home.

Younger than most people who gradually descend into the darkness of Alzheimer's, Hart was able to express her loneliness, her fears, and her difficulties coping. She was about the same age as Auguste Deter, Alois Alzheimer's first patient with the disease that now bears his name. In 1901, when Alzheimer examined Mrs. Deter upon her admission to the city mental asylum in Frankfurt-am-Main, he found a disoriented and agitated woman who had difficulty finding words. She was confused, anxious, and reluctant to cooperate with the staff. She exhibited difficulties in both reading and writing. Alzheimer was impressed by her impaired memory: When objects were shown to her, she was unable to recall them after a short interval; moreover, she was unable to write down a short sentence after it was dictated to her even though she could write the words when they were dictated one after another. Neurological examination showed her gait, pupils, and the motor functions of her extremities were normal. During hospitalization she became progressively more apathetic, spent much of the time in bed in a fetal position, and was incontinent. She died five years after admission, and her brain was removed at autopsy. Alzheimer saw that the brain was shrunken; after cutting sections and staining them with a silver preparation, he found numerous plaques and tangles under the microscope.

For almost seventy-five years, Alzheimer's disease was classified as a presenile dementia, afflicting people under sixty. The dementia, described by Alzheimer in 1906, was considered distinct from so-called senile dementia—or senility for short—which became progressively more common as life span increased. Even though the majority of elderly people have well-preserved memories and intellect, senile dementia was thought to be an inevitable consequence of aging. In the mid-1970s, medical scientists began to argue that the plaques and tangles in the brains of patients with "senile dementia" were indistinguishable from those found in Alzheimer's, and that we should label all patients with these pathologic changes as Alzheimer's regardless of their age (see figure 37).[6] Senility, in other words, is not a "normal" consequence of aging. The plaques and tangles first reported by Alois Alzheimer remained enigmatic until the mid-1980s, when the major protein components of these structures were identified.[7]

Over the past two hundred years, life expectancy of Americans has tripled—from about twenty-five years in colonial times to more than seventy-five currently. During the past fourscore years, one remarkable discovery after another has contributed to this tripling: Antibiotics and antivirals cured numberless infections; vaccines eliminated polio, measles, mumps; antihypertensive drugs and anticoagulants prevented heart attacks and strokes; chemotherapy and radiotherapy cured many cancers. With new diagnostic procedures and innovative therapeutic interventions, the life expectancy of Americans could reach one hundred by the middle of this century.[8]

And yet . . . half of all Americans suffer from a nervous system disorder at some time in their lives, and these maladies—schizophrenia, depression, stroke, Alzheimer's, Parkinson's, brain tumors, meningitis, epilepsy, headache, deafness, impaired vision, autism, traumatic brain injury—are often chronic and debilitating. For some of these illnesses, we have medicines that diminish the symptoms, but they rarely halt the underlying pathogenic process. For example, the tremor and rigidity of Parkinson's can be treated effectively with L-dopa early in the disease, but the underlying destruction of neurons continues unabated.[9] More than 1.5 million Americans suffer the ravages of Parkinson's.

The prevalence of both Parkinson's and Alzheimer's combined is about seven million in the United States, and both of these illnesses increase dramatically with age. So if we extend the life span of Americans without confronting this onslaught, we will, in effect, be creating a nation of demented, frozen people. Alzheimer's is rare before age sixty but afflicts nearly one in twenty by age seventy and one in three by eighty-five—so if you escape cancer, heart disease, and stroke, and live to the age of eighty-five, you have a 35 percent chance of developing Alzheimer's.[10] Those are terrible odds. Even a modest increase in life expectancy over the next forty years will triple the number of people in the United States with Alzheimer's—and that means huge expenditures for their care and untold suffering for patients and their families.

While we can quantify patient numbers and ages, we cannot quantify the fear and hopelessness that Alzheimer's patients experience as they steadily lose their intellect in the early phases of disease. And we cannot measure the suffering felt by the spouses, children, and close friends of Alzheimer's victims—they are powerless to slow the deterioration of their loved ones. They watch and wait, helpless to change the course of this dreaded illness.

The New Psychiatry

What about younger people who develop neurodegeneration? Are we to ignore the soldiers who defend our freedom and the athletes who bring so much entertainment into our lives? Will the parents of soldiers and athletes stand by and let their children suffer from the cruel neuropsychiatric illnesses arising from the accumulation of tau prions? Like Alzheimer's victims, former soldiers and athletes suffering from the ravages of posttraumatic FTD are unable to demand research directed toward curing their diseases.

At present, hundreds of thousands of U.S. veterans carry the diagnosis of posttraumatic stress disorder: more than 800,000 from the

Vietnam War and more than 300,000 from the Iraq and Afghanistan wars. Many active-duty combatants and veterans who committed suicide were diagnosed with posttraumatic stress disorder. The term "posttraumatic stress disorder" (PTSD) represents an array of behavioral problems that include severe anxiety, nightmares, apathy, hypervigilance, depression, insomnia, and diminished executive function. Some patients become socially inappropriate and withdrawn; some become addicted to drugs as well as alcohol. Many patients with posttraumatic stress disorder experience an escalation of the severity of their symptoms over time, and some show progressive deterioration of brain function.

How many of the veterans suffering from posttraumatic stress disorder have a posttraumatic form of FTD is unknown. Studies of U.S. combatants argue that mild concussions significantly increase the likelihood of developing posttraumatic stress disorder.[11] How many military personnel with posttraumatic stress disorder have tau prions, induced by head trauma, proliferating in their brains remains to be determined. Brain-imaging procedures that can detect tau prions will have to be developed in order to answer such questions.

Leveraging the Prion Revolution

It is time to launch a bold research plan to fathom the inner workings if the brain. We have the knowledge required to design an unprecedented voyage of discovery in the neurosciences, so that we can begin to understand the very essence of being human. Such a scientific research agenda must have the same national priority that we gave to two remarkable engineering projects: building an atomic bomb and sending humans to the moon.

I am confident that with the necessary research, scientists can discover in the not-too-distant future how we think, remember, dream, reason, imagine, and adapt. They will also decipher the molecular basis of consciousness, which will lead to a better understanding of such complex feelings as happiness, well-being, joy, discontent, depression, anger, and sadness. Elucidating the chemistry of learning and memory will lead to the discovery of new drug targets for many nervous system ailments.

If developed nations devote sufficient resources, I can anticipate a time when people will no longer suffer from the ravages of Alzheimer's and Parkinson's diseases or the FTDs. The only issue is how many more people will die from these illnesses before cures are created. The time frame will continue to be governed largely by the size of the scientific effort. Unfortunately, demented patients cannot function as their own advocates. And in my experience, few spouses and children of Alzheimer's victims become strong advocates for or generous donors to neurodegenerative-disease research programs. As modern medicine increases the life span of people worldwide, the number of older people developing neurodegenerative diseases will continue to rise. It seems to me that developing meaningful preventions and effective cures for Alzheimer's and Parkinson's should be among the highest of our national priorities.

The prion revolution, which now encompasses Alzheimer's, Parkinson's, the FTDs, ALS, and Huntington's disease, has provided new insights into the pathogenesis of these age-dependent, devastating illnesses.[12] The new vistas that have been opened by investigations demonstrating that prions cause these neurodegenerative disorders promise to enhance our understanding and facilitate research. Once so mysterious, degenerative brain diseases are beginning to reveal the mechanisms by which they compromise nervous system function.

Drug discovery studies for CJD, Alzheimer's, Parkinson's, and the FTDs have been too few. That our nation has ignored these devastating illnesses of older people is an inescapable conclusion—witness the NIH budget. For the wars on cancer and heart disease, there are a myriad of advocates and pipelines filled with hundreds of promising drugs—for neurodegenerative diseases, the ranks and the pipelines are virtually empty.

Notes

Preface

1. T. S. Kuhn, *The Structure of Scientific Revolutions* (University of Chicago Press, Chicago, 1962); C. M. Christensen, *The Innovator's Dilemma: When New Technologies Cause Great Firms to Fail* (Harvard Business School Press, Boston, 1997).

2. N. Bilton, "Disruptions: Innovation isn't easy, especially midstream," *New York Times,* April 15, 2012.

3. S. B. Prusiner, "Novel proteinaceous infectious particles cause scrapie," *Science* 216: 136–44 (1982).

4. S. B. Prusiner, "A unifying role for prions in neurodegenerative diseases," *Science* 336: 1511–13 (2012).

Introduction

1. G. M. Shy and G. A. Drager, "A neurological syndrome associated with orthostatic hypotension: A clinical-pathologic study," *Arch. Neurol.* 2: 511–27 (1960).

2. W. K. Engel, "A further fond farewell to Shy and Drager," *Ann. Intern. Med.* 126: 334 (1997); I. J. Schatz, "Farewell to the 'Shy-Drager syndrome,'" *Ann. Intern. Med.* 125: 74–75 (1996).

3. T. C. Chamberlin, "The method of multiple working hypotheses," *Science* [Old Series] 15: 92–97 (1890).

1. Growing Up

1. S. Prusiner and S. K. Wolfson, Jr., "Hypothermic protection against cerebral edema of ischemia," *Arch. Neurol.* 19: 623–27 (1968).

2. B. Black-Schaffer, S. Prusiner, and H. Esparza, "Tolerance of the vestibular apparatus of the hypothermic hamster to 840G acceleration," *Aerospace Med.* 36: 123–26 (1965).

3. S. Prusiner, J. R. Williamson, B. Chance, and B. M. Paddle, "Pyridine nucleotide changes during thermogenesis in brown fat tissue *in vivo,*" *Arch. Biochem. Biophys.* 123: 368–77 (1968).

2. The Beginning of an Odyssey

1. C. J. Gibbs, Jr., D. C. Gajdusek, D. M. Asher, M. P. Alpers, E. Beck, P. M. Daniel, and W. B. Matthews, "Creutzfeldt-Jakob disease (spongiform encephalopathy): Transmission to the chimpanzee," *Science* 161: 388–89 (1968).

2. D. C. Gajdusek, C. J. Gibbs, Jr., and M. Alpers, "Experimental transmission of a kuru-like syndrome to chimpanzees," *Nature* 209: 794–96 (1966).

3. See esp. T. Alper, W. A. Cramp, D. A. Haig, and M. C. Clarke, "Does the agent of scrapie replicate without nucleic acid?" *Nature* 214: 764–66 (1967).

4. Carl Sagan, an astronomer always interested in prebiotic chemicals and the origins of life on Earth, wrote me inquiring whether I thought the scrapie agent might have arrived here from outer space, inside or on the surface of a meteorite. He wondered whether the extreme resistance to radiation found by Alper might have protected it during its long journey. I replied that I was doubtful, but that we needed to purify the scrapie agent and determine its composition before I could answer his provocative query.

5. University of California, San Francisco, "Interviews with Robert A. Fishman, M.D.: Building Neurology, Teaching, Research, and Patient Care," in the UCSF Oral History Program, Department of the History of Health Sciences (Regents of the University of California, San Francisco, 1998).

3. A Plethora of Theories

1. R. L. Chandler, "Encephalopathy in mice produced by inoculation with scrapie brain material," *Lancet* 1, no. 719: 1378–79 (1961).

2. T. Alper, D. A. Haig, and M. C. Clarke, "The exceptionally small size of the scrapie agent," *Biochem. Biophys. Res. Commun.* 22: 278–84 (1966).

3. R. Latarjet, B. Muel, D. A. Haig, M. C. Clarke, and T. Alper, "Inactivation of the scrapie agent by near monochromatic ultraviolet light," *Nature* 227: 1341–43 (1970).

4. J. S. Griffith, "Nature of the scrapie agent: Self-replication and scrapie," *Nature* 215: 1043–44 (1967).

5. D. C. Gajdusek, "Unconventional viruses and the origin and disappearance of kuru," *Science:* 197: 943–60 (1977).

6. I. H. Pattison, "Fifty years with scrapie: A personal reminiscence," *Vet. Rec.* 123: 661–66 (1988).

7. W. J. Hadlow, "Scrapie and kuru," *Lancet* 2, no. 289–90 (1959).

8. When I later asked him why he hadn't embarked on such a project himself (after all, he had had training in human neuropathology and as a veterinarian he was qualified to perform the chimpanzee experiments later done by Gibbs and Gajdusek), Hadlow shrugged off the question, professing to

have been more interested in returning to Montana to study scrapie than in pursuing kuru.

4. The Scrapie Race

1. S. B. Prusiner, "An approach to the isolation of biological particles using sedimentation analysis," *J. Biol. Chem.* 253: 916–21 (1978).

2. S. B. Prusiner, W. J. Hadlow, C. M. Eklund, and R. E. Race, "Sedimentation properties of the scrapie agent," *Proc. Natl. Acad. Sci. USA* 74: 4656–60 (1977).

3. M. A. Chatigny, K. Ishimaru, S. Dunn, J. A. Eagleson, and S. B. Prusiner, "Evaluation of a class III biological safety cabinet for enclosure of an ultracentrifuge," *Appl. Environ. Microbiol.* 38: 934–39 (1979).

4. G. D. Hunter, *Scrapie and Mad Cow Disease: The Smallest and Most Lethal Living Thing* (Vantage, New York, 1993), 1–115; G. D. Hunter, "The enigma of the scrapie agent: Biochemical approaches and the involvement of membranes and nucleic acids," in *Slow Transmissible Diseases of the Nervous System,* vol. 2, ed. S. B. Prusiner and W. J. Hadlow (Academic Press, New York, 1979), 365–85.

5. G. D. Hunter, R. H. Kimberlin, G. C. Millson, and R. A. Gibbons, "An experimental examination of the scrapie agent in cell membrane mixtures. I. Stability and physicochemical properties of the scrapie agent," *J. Comp. Pathol.* 81: 23–32 (1971); G. C. Millson, G. D. Hunter, and R. H. Kimberlin, "An experimental examination of the scrapie agent in cell membrane mixtures. II. The association of scrapie infectivity with membrane fractions," *J. Comp. Pathol.* 81: 255–65 (1971); R. H. Kimberlin, G. C. Millson, and G. D. Hunter, "An experimental examination of the scrapie agent in cell membrane mixtures. III. Studies of the operational size," *J. Comp. Pathol.* 81: 383–91 (1971).

6. R. G. Rohwer and D. C. Gajdusek, "Scrapie—virus or viroid: The case for a virus," in *Search for the Cause of Multiple Sclerosis and Other Chronic Diseases of the Central Nervous System,* ed. A. Boese (Verlag Chemie, Weinheim, 1980), 333–55.

7. H. J. Cho, A. S. Grieg, C. R. Corp, R. H. Kimberlin, R. L. Chandler, and G. C. Millson, "Virus-like particles from both control and scrapie-affected mouse brain," *Nature* 267: 459–60 (1977).

8. J. F. David-Ferreira, K. L. David-Ferreira, C. J. Gibbs, Jr., and J. A. Morris, "Scrapie in mice: Ultrastructural observations in the cerebral cortex," *Proc. Soc. Exp. Biol. Med.* 127: 313–20 (1968).

9. J. R. Baringer and S. B. Prusiner, "Experimental scrapie in mice: Ultrastructural observations," *Ann. Neurol.* 4: 205–11 (1978).

10. A. G. Dickinson and G. W. Outram, "An assessment of the genetics of scrapie in sheep and mice," in *Slow Transmissible Diseases of the Nervous System,* ed. S. B. Prusiner and W. J. Hadlow (Academic Press, New York, 1979), 1: 367–86.

11. R. H. Kimberlin, "Scrapie agent: Prions or virinos?" *Nature* 297: 107–8 (1982).

12. R. F. Marsh, T. G. Malone, J. S. Semancik, W. D. Lancaster, and R. P. Hanson, "Evidence for an essential DNA component in the scrapie agent," *Nature* 275: 146–47 (1978).

13. S. B. Prusiner, M. P. McKinley, D. F. Groth, K. A. Bowman, N. I. Mock, S. P. Cochran, and F. R. Masiarz, "Scrapie agent contains a hydrophobic protein," *Proc. Natl. Acad. Sci. USA* 78: 6675–79 (1981).

5. Dr. America and the Trembling Cannibals

1. V. Zigas and D. C. Gajdusek, "Kuru: Clinical study of a new syndrome resembling *paralysis agitans* in natives of the Eastern Highlands of Australian New Guinea," *Med. J. Australia* 44: 745–54 (1957).

2. M. P. Alpers, "Kuru: Implications of its transmissibility for the interpretation of its changing epidemiological pattern," in *The Central Nervous System: Some Experimental Models of Neurological Diseases,* ed. O. T. Bailey and D. E. Smith (Williams and Wilkins, Baltimore, 1968), 234–51.

3. R. Glasse, "Cannibalism in the kuru region of New Guinea," *Trans. N.Y. Acad. Sci.* [Series 2] 29: 748–54 (1967).

4. See esp. S. Lindenbaum, "Cannibalism, kuru, and anthropology," *Folia Neuropathol.* 47: 138–44 (2009).

5. C. J. Gibbs, Jr., H. L. Amyx, A. Bacote, C. L. Masters, and D. C. Gajdusek, "Oral transmission of kuru, Creutzfeldt-Jakob disease, and scrapie to nonhuman primates," *J. Infect. Dis.* 142: 205–8 (1980).

6. A. Jakob, "Über eigenartige Erkrankungen des Zentralnervensystems mit bemerkenswertem anatomischem Befunde (Spastische Pseudosklerose-Encephalomyelopathie mit disseminierten Degenerationsherden)," *Z. Gesamte Neurol. Psychiatr.* 64: 147–228 (1921).

7. H. G. Creutzfeldt, "Über eine eigenartige herdförmige Erkrankung des Zentralnervensystems," *Z. Gesamte Neurol. Psychiatr.* 57: 1–18 (1920).

8. C. J. Gibbs, Jr., "Spongiform encephalopathies—slow, latent, and temperate virus infections—in retrospect," in *Prion Diseases of Humans and Animals,* ed. S. B. Prusiner, J. Collinge, J. Powell, and B. Anderton (Ellis Horwood, London, 1992), 53–62.

9. R. W. Hornabrook, "Kuru—a subacute cerebellar degeneration: The natural history and clinical features," *Brain* 91: 53–74 (1968).

10. B. Lindquist, writer–director, "The Genius and the Boys," *Storyville,* exec. prod. J. Kellagher (London: BBC Four, 2009).

6. The Battle for Tenure

1. R. H. Kimberlin and R. F. Marsh, "Comparison of scrapie and transmissible mink encephalopathy in hamsters. I. Biochemical studies of brain during development of disease," *J. Infect. Dis.* 131: 97–103 (1975); R. F. Marsh and R. H. Kimberlin, "Comparison of scrapie and transmissible mink encephalopathy in hamsters. II. Clinical signs, pathology, and pathogenesis," *J. Infect. Dis.* 131: 104–10 (1975).

2. University of California, San Francisco, "Interviews with Robert A. Fishman, M.D.: Building Neurology, Teaching, Research, and Patient Care," in the UCSF Oral History Program, Department of the History of Health Sciences (Regents of the University of California, San Francisco, 1998).

7. What's in a Name?

1. To some of our critics, this unusual purification technique sounded just as crazy as the idea of an infectious protein itself. But among all the scientists working in the field, I probably had the most experience purifying proteins. In Earl Stadtman's lab at the NIH, I had learned how to purify proteins found in low abundance. While I had no experience isolating viruses, in the end I didn't need it. That the scrapie agent turned out to be a protein was another stroke of luck.

2. S. B. Prusiner, "Novel proteinaceous infectious particles cause scrapie," *Science* 216: 136–44 (1982).

3. R. H. Kimberlin, "Scrapie agent: Prions or virinos?" *Nature* 297: 107–8 (1982).

4. C. J. Gibbs, Jr., D. C. Gajdusek, and R. Latarjet, "Unusual resistance to ionizing radiation of the viruses of kuru, Creutzfeldt-Jakob disease, and scrapie," *Proc. Nat. Acad. Sci. USA* 75: 6268–70 (1978).

5. R. T. Johnson, "The novel nature of scrapie." *Trends Neurosci.* 5: 413–15 (1982).

6. T. O. Diener, M. P. McKinley, and S. B. Prusiner, "Viroids and prions," *Proc. Nat. Acad. Sci. USA* 79: 5220–24 (1982).

7. O. T. Avery, C. M. MacLeod, and M. McCarty, "Studies on the chemical nature of the substance inducing transformation of pneumococcal types. Induction of transformation by a deoxyribonucleic acid fraction isolated from pneumococcus type III," *J. Exp. Med.* 79: 137–57 (1944).

8. Lost in the Pacific Fog

1. D. C. Bolton, M. P. McKinley, and S. B. Prusiner, "Identification of a protein that purifies with the scrapie prion," *Science* 218: 1309–11 (1982).

2. S. B. Prusiner, D. C. Bolton, D. F. Groth, K. A. Bowman, S. P. Cochran, and M. P. McKinley, "Further purification and characterization of scrapie prions," *Biochemistry* 21: 6942–50 (1982).

3. M. P. McKinley, D. C. Bolton, and S. B. Prusiner, "A protease-resistant protein is a structural component of the scrapie prion," *Cell* 35: 57–62 (1983).

4. Several years before I entered the field, Hilary Koprowski, at the Wistar Institute in Philadelphia, had looked in vain for activation of the interferon system in mice inoculated with the scrapie agent; M. Katz and H. Koprowski, "Failure to demonstrate a relationship between scrapie and production of interferon in mice," *Nature* 219: 639–40 (1968).

5. L. K. Altman, "The doctor's world: The mystery of Balanchine's death is solved," *New York Times,* May 8, 1984.

9. The Amyloid Story

1. S. B. Prusiner, D. C. Bolton, D. F. Groth, K. A. Bowman, S. P. Cochran, and M. P. McKinley, "Further purification and characterization of scrapie prions," *Biochemistry* 21: 6942–50 (1982).

2. A. S. Cohen, T. Shirahama, and M. Skinner, "Electron microscopy of amyloid," in *The Electron Microscopy of Proteins,* vol. 3, ed. J. R. Harris (Academic Press, New York, 1982), 165–206.

3. P. Divry, "Etude histochimique des plaques seniles," *J. Belge Neurol. Psychiatrie* 27: 643–54 (1927); I. Klatzo, D. C. Gajdusek, and V. Zigas, "Pathology of kuru," *Lab. Invest.* 8: 799–847 (1959).

4. S. B. Prusiner, M. P. McKinley, K. A. Bowman, D. C. Bolton, P. E. Bendheim, D. F. Groth, and G. G. Glenner, "Scrapie prions aggregate to form amyloid-like birefringent rods," *Cell* 35: 349–58 (1983).

5. L. K. Altman, "Substance tied to Alzheimer's in Coast study," *New York Times,* December 7, 1983.

6. G. Taubes, "The name of the game is fame. But is it science?" *Discover* 7: 28–52 (1986).

7. G. G. Glenner and C. W. Wong, "Alzheimer's disease: Initial report of the purification and characterization of a novel cerebrovascular amyloid protein," *Biochem. Biophys. Res. Commun.* 120: 885–90 (1984).

8. S. B. Prusiner, "Some speculations about prions, amyloid, and Alzheimer's disease," *N. Engl. J. Med.* 310: 661–63 (1984).

9. P. E. Bendheim, R. A. Barry, S. J. DeArmond, D. P. Stites, and S. B. Prusiner, "Antibodies to a scrapie prion protein," *Nature* 310: 418–21 (1984).

10. S. J. DeArmond, M. P. McKinley, R. A. Barry, M. B. Braunfeld, J. R. McColloch, and S. B. Prusiner, "Identification of prion amyloid filaments in scrapie-infected brain," *Cell* 41: 221–35 (1985).

11. S. J. DeArmond, H. A. Kretzschmar, M. P. McKinley, and S. B. Prusiner, "Molecular pathology of prion diseases," in *Prions: Novel Infectious Pathogens Causing Scrapie and Creutzfeldt-Jakob Disease,* ed. S. B. Prusiner and M. P. McKinley (Academic Press, Orlando, Fla., 1987) 387–414.

12. P. A. Merz, H. M. Wisniewski, R. A. Somerville, S. A. Bobin, C. L. Masters, and K. Iqbal, "Ultrastructural morphology of amyloid fibrils from neuritic and amyloid plaques," *Acta Neuropathol. (Berl.)* 60: 113–24 (1983).

13. H. Diringer, H. Gelderblom, H. Hilmert, M. Ozel, C. Edelbluth, and R. H. Kimberlin, "Scrapie infectivity, fibrils and low molecular weight protein," *Nature* 306: 476–78 (1983).

14. P. A. Merz, R. G. Rohwer, R. Kascsak, H. M. Wisniewski, R. A. Somerville, C. J. Gibbs, Jr., and D. C. Gajdusek, "Infection-specific particle from the unconventional slow virus diseases," *Science* 225: 437–40 (1984).

15. Diringer was not alone. Gajdusek was to write, a half-dozen years later (D. C. Gajdusek, "Subacute spongiform encephalopathies: Transmissible cerebral amyloidoses caused by unconventional viruses," in *Virology,* 2nd ed., ed. B. N. Fields, D. M. Knipe, R. M. Chanock, M. S. Hirsch, J. L. Melnick, T. P. Monath, and B. Roizman [Raven, New York, 1990], 2289–324): "In the transmissible brain amyloidoses caused by the unconventional viruses, the process appears to be activated by a virus that nucleates the initiation of the structural change of the normal precursor protein to the β-pleated insoluble and protease-resistant configuration of the amyloid subunit. Oligomeric assemblies of these subunits probably nucleate their own fibril polymerization into SAFs and the fibrils of the amyloid plaques in these diseases. These microfibrils condense by crystallization into the microscopically visible amyloid structures we appreciate." Like Diringer, he persisted in thinking that atypical viruses causing scrapie, CJD, and kuru stimulated the formation of PrP 27–30, which then polymerized into amyloid fibrils.

10. Finding the Gene

1. S. B. Prusiner, "Novel proteinaceous infectious particles cause scrapie," *Science* 216: 136–44 (1982).

2. A. Koglin and C. T. Walsh, "Structural insights into nonribosomal peptide enzymatic assembly lines," *Nat. Prod. Rep.* 26: 987–1000 (2009).

3. S. B. Prusiner, D. F. Groth, D. C. Bolton, S. B. Kent, and L. E. Hood, "Purification and structural studies of a major scrapie prion protein," *Cell* 38: 127–34 (1984).

4. At the Montana meeting, Diringer read from my one of my papers recently published in *Biochemistry* and claimed that I had said that the scrapie agent consisted only of a protein. I had not said that, and several members of the audience informed Diringer that he had misinterpreted my English. The episode opened my eyes; I learned how important every word in a scientific manuscript can be.

5. Shortly after we met, I learned that Erling had been involved in the decision to give the 1976 Nobel Prize in physiology or medicine to Baruch Bloomberg and Carleton Gajdusek, and I asked him why Bill Hadlow and Joe Gibbs, or Michael Alpers and Gibbs, or Alpers and Hadlow, had not shared the prize with Gajdusek. Erling closed the conversation gracefully; he could not reveal any of the deliberations until 2026, when the papers relating to the 1976 prize will be open to scholars. I still don't know whether Hadlow or Alpers or Gibbs was ever nominated, or why it was decided to award half the prize to Gajdusek and half to someone in another field (hepatitis B virus).

6. B. Oesch, D. Westaway, M. Wälchli, M. P. McKinley, S. B. H. Kent, R. Aebersold, R. A. Barry, P. Tempst, D. B. Teplow, L. E. Hood, S. B. Prusiner, and C. Weissmann, "A cellular gene encodes scrapie PrP 27–30 protein," *Cell* 40: 735–46 (1985).

7. B. Chesebro, R. Race, K. Wehrly, J. Nishio, M. Bloom, D. Lechner, S. Bergstrom, K. Robbins, L. Mayer, J. M. Keith, C. Garon, and A. Haase, "Identification of scrapie prion protein-specific mRNA in scrapie-infected and uninfected brain," *Nature* 315: 331–33 (1985).

8. D. T. Kingsbury, K. C. Kasper, D. P. Stites, J. D. Watson, R. N. Hogan, and S. B. Prusiner, "Genetic control of scrapie and Creutzfeldt-Jakob disease in mice," *J. Immunol.* 131: 491–96 (1983).

9. A. G. Dickinson, V. M. H. Meikle, and H. Fraser, "Identification of a gene which controls the incubation period of some strains of scrapie agent in mice," *J. Comp. Pathol.* 78: 293–99 (1968).

10. Eventually we learned that prions in I/LnJ mice multiplied much slower than those in NZW mice because of a two–amino acid difference in the sequence of the precursor protein PrP^C that David Westaway had found earlier. The slower production of new prions resulted in prolonged incubation times.

11. Prompted by a suggestion from Berkeley virologist David Kingsbury. G. A. Carlson, "Prions and prejudice: From *Sinc* to systems with Stan," in *Prion*

Research of Stan Prusiner and His Colleagues, ed. G. Legname and D. Riesner (Düsseldorf University Press, Düsseldorf, 2013), 47–61.

12. G. A. Carlson, D. T. Kingsbury, P. A. Goodman, S. Coleman, S. T. Marshall, S. DeArmond, D. Westaway, and S. B. Prusiner, "Linkage of prion protein and scrapie incubation time genes," *Cell* 46: 503–11 (1986).

13. D. Westaway, P. A. Goodman, C. A. Mirenda, M. P. McKinley, G. A. Carlson, and S. B. Prusiner, "Distinct prion proteins in short and long scrapie incubation period mice," *Cell* 51: 651–62 (1987).

14. N. Hunter, J. Hope, I. McConnell, and A. G. Dickinson, "Linkage of the scrapie-associated fibril protein (PrP) gene and *Sinc* using congenic mice and restriction fragment length polymorphism analysis," *J. Gen. Virol.* 68: 2711–16 (1987).

11. Jousting with the Press

1. Despite the following account of my difficulties with the press, I want to mention that every article written by David Perlman of the *San Francisco Chronicle* over nearly three decades about my work has been fair, accurate, and balanced. The same can be said of Larry Altman and Sandra Blakeslee at the *New York Times.*

2. F. Peterson and J. Kesselman-Turkel, "Agent X," *Omni* 4: 96–100 (1981).

3. G. Taubes, "The name of the game is fame. But is it science?" *Discover* 7: 28–52 (1986).

4. S. B. Prusiner, "Novel proteinaceous infectious particles cause scrapie," *Science* 216: 136–44 (1982).

5. C. Weissmann, "Prusiner and prions," *Discover* 8: 102 (1987).

6. Howard Temin proposed the idea of proviruses to explain how RNA tumor viruses seem to disappear after entering cells and then reappear later. Many people thought his interpretation of the findings was a big mistake. All that changed when, independently, David Baltimore and Temin identified an enzyme that copies RNA into DNA. The enzyme, called reverse transcriptase, created a DNA copy of the RNA virus genome; this DNA copy was the provirus Temin had hypothesized many years earlier. In 1975, Baltimore and Temin were awarded the Nobel Prize for their work.

12. Deciphering Human Prion Diseases

1. R. Roos, D. C. Gajdusek, and C. J. Gibbs, Jr., "The clinical characteristics of transmissible Creutzfeldt-Jakob disease," *Brain* 96: 1–20 (1973).

2. C. L. Masters, D. C. Gajdusek, and C. J. Gibbs, Jr., "Creutzfeldt-Jakob disease virus isolations from the Gerstmann-Sträussler syndrome," *Brain* 104: 559–88 (1981).

3. J. Gerstmann, E. Sträussler, and I. Scheinker, "Über eine eigenartige hereditär-familiäre Erkrankung des Zentralnervensystems zugleich ein Beitrag zur frage des vorzeitigen lokalen Alterns," *Z. Neurol.* 154: 736–62 (1936). When Hitler annexed Austria in March 1938, the university positions of all Jewish academics were terminated. Both Gerstmann and Scheinker left Austria and settled in the United States. Sträussler decided to stay in Vienna, since he was married to a Catholic woman and they had one son, whom they raised as a Catholic; according to the Nuremberg race law on *Mischehen* (mixed marriages), Sträussler was classified as "privileged." Although he could no longer work, he received a pension, was not forced to wear a yellow Star of David, and was not sent to a death camp. He worked as a forensic psychiatrist after the war ended, living in Vienna until his death in 1959. Gerstmann had a distinguished career in neurology and psychiatry, initially in Washington, D.C., and later in New York; he died in 1969 at age eighty-two. Scheinker worked as a neuropathologist, first at the University of Cincinnati and later in New York, and died of a heart attack at fifty-two. Unfortunately, none of the three lived to see the impact that their clinical studies would have on the development of a new paradigm of disease.

4. J. A. Hainfellner, S. Brantner-Inthaler, L. Cervenáková, P. Brown, T. Kitamoto, J. Tateishi, H. Diringer, P. P. Liberski, H. Regele, M. Feucht, N. Mayr, P. Wessely, K. Summer, F. Seitelberger, and H. Budka, "The original Gerstmann-Sträussler-Scheinker family of Austria: divergent clinicopathological phenotypes but constant PrP genotype," *Brain Pathol.* 5: 201–11 (1995).

5. The UCLA neuropathologist was Dr. Harry Vinters.

6. All codons comprise three of the four bases—A, T, G, and C—and direct the amino acid to be assembled; the codon number refers to the position of the encoded amino acid.

7. K. Hsiao, H. F. Baker, T. J. Crow, M. Poulter, F. Owen, J. D. Terwilliger, D. Westaway, J. Ott, and S. B. Prusiner, "Linkage of a prion protein missense variant to Gerstmann-Sträussler syndrome," *Nature* 338: 342–45 (1989).

8. J. Safar, M. Ceroni, P. Piccardo, P. P. Liberski, M. Miyazaki, D. C. Gajdusek, and C. J. Gibbs, Jr., "Subcellular distribution and physicochemical properties of scrapie-associated precursor protein and relationship with scrapie agent," *Neurology* 40: 503–8 (1990); M. Ceroni, P. Piccardo, J. Safar, D. C. Gajdusek, and C. J. Gibbs, Jr., "Scrapie infectivity and prion protein are distributed in the same pH range in agarose isoelectric focusing," *Neurology* 40: 508–13 (1990); J. Safar, M. Ceroni, P. Piccardo, D. C. Gajdusek, and C. J. Gibbs, Jr., "Scrapie-associated precursor proteins: Antigenic relationship

between species and immunocytochemical localization in normal, scrapie, and Creutzfeldt-Jakob disease brains," *Neurology* 40: 513–17 (1990); P. Piccardo, J. Safar, M. Ceroni, D. C. Gajdusek, and C. J. Gibbs, Jr., "Immunohistochemical localization of prion protein in spongiform encephalopathies and normal brain tissue," *Neurology* 40: 518–22 (1990).

9. M. Scott, D. Foster, C. Mirenda, D. Serban, F. Coufal, M. Wälchli, M. Torchia, D. Groth, G. Carlson, S. J. DeArmond, D. Westaway, and S. B. Prusiner, "Transgenic mice expressing hamster prion protein produce species-specific scrapie infectivity and amyloid plaques," *Cell* 59: 847–57 (1989).

10. Later, Mike Scott managed to reduce the number of possible amino acids controlling the prion replication process from eleven to five by producing mice expressing a PrP transgene that was part hamster and part mouse.

11. M. Scott, D. Foster, C. Mirenda, D. Serban, F. Coufal, M. Wälchli, M. Torchia, D. Groth, G. Carlson, S. J. DeArmond, D. Westaway, and S. B. Prusiner, "Transgenic mice expressing hamster prion protein produce species-specific scrapie infectivity and amyloid plaques," *Cell* 59: 847–57 (1989).

12. C. Weissmann, "A unified theory of prion propagation," *Nature* 352: 679–83 (1991).

13. R. Medori, H.-J. Tritschler, A. LeBlanc, F. Villare, V. Manetto, H. Y. Chen, R. Xue, S. Leal, P. Montagna, P. Cortelli, P. Tinuper, P. Avoni, M. Mochi, A. Baruzzi, J. J. Hauw, J. Ott, E. Lugaresi, L. Autilio-Gambetti, and P. Gambetti, "Fatal familial insomnia, a prion disease with a mutation at codon 178 of the prion protein gene," *N. Engl. J. Med.* 326: 444–49 (1992).

14. F. Owen, M. Poulter, J. Collinge, and T. J. Crow, "Codon 129 changes in the prion protein gene in Caucasians," *Am. J. Hum. Genet.* 46: 1215–16 (1990).

13. What's in a Shape?

1. B. W. Caughey, A. Dong, K. S. Bhat, D. Ernst, S. F. Hayes, and W. S. Caughey, "Secondary structure analysis of the scrapie-associated protein PrP 27–30 in water by infrared spectroscopy," *Biochemistry* 30: 7672–80 (1991).

2. K.-M. Pan, N. Stahl, and S. B. Prusiner, "Purification and properties of the cellular prion protein from Syrian hamster brain," *Protein Sci.* 1: 1343–52 (1992); K.-M. Pan, M. Baldwin, J. Nguyen, M. Gasset, A. Serban, D. Groth, I. Mehlhorn, Z. Huang, R. J. Fletterick, F. E. Cohen, and S. B. Prusiner, "Conversion of α-helices into β-sheets features in the formation of the scrapie prion proteins," *Proc. Natl. Acad. Sci. USA* 90: 10962–66 (1993).

3. C. Weissmann, H. Büeler, M. Aguet, M. Fischer, and A. Sailer. "Transgenic animals lacking prion proteins." United States #5,698,763; 1997.

4. H. Büeler, M. Fisher, Y. Lang, H. Bluethmann, H.-P. Lipp, S. J. DeArmond, S. B. Prusiner, M. Aguet, and C. Weissmann, "Normal development and

behaviour of mice lacking the neuronal cell-surface PrP protein," *Nature* 356: 577–82 (1992).

14. Turf Battles

1. S. B. Prusiner and M. McCarty, "Discovering DNA encodes heredity and prions are infectious proteins," *Annu. Rev. Genet.* 40: 25–45 (2006).

2. G. Kolata, "Viruses or prions: An old medical debate still rages," *New York Times*, October 4, 1994.

3. D. C. Gajdusek, "Unconventional viruses causing subacute spongiform encephalopathies," in *Virology,* ed. B. N. Fields, D. M. Knipe, R. M. Chanock, J. L. Melnick, B. Roizman, and R. E. Shope (Raven, New York, 1985) 1519–57. My first recollection of the Gajdusek problem dates to 1982, when he gave the prestigious Hitchcock Lectures at UC Berkeley. One afternoon, Robley Williams telephoned me. "Stan," he said, "have you ever met a man who doesn't need oxygen? I have had Carleton Gajdusek in my office for the last two hours, and he never stopped exhaling. Not once did he stop and take a breath." Gajdusek had treated him to a disquisition on my work, had claimed it was his, and—it seemed to Robley—had misinterpreted much of it.

4. R. G. Rohwer and D. C. Gajdusek, "Scrapie—virus or viroid: The case for a virus," in *Search for the Cause of Multiple Sclerosis and Other Chronic Diseases of the Central Nervous System,* ed. A. Boese (Verlag Chemie, Weinheim, 1980), 333–55.

5. I. Klatzo, D. C. Gajdusek, and V. Zigas, "Pathology of kuru," *Lab. Invest.* 8: 799–847 (1959).

6. Ephraim was also an Israeli hero. While working in the Haganah (the Jewish underground), Ephraim and his older brother, Aharon, developed new explosives that were used in the War of Independence in 1948. Later he would serve as chief scientist for the Israeli Defense Ministry. A modest man, Ephraim came into public life when Prime Minister Golda Meir asked him to stand for election for the presidency of Israel in the spring of 1973. Her first choice, his brother, Aharon, had been killed by Japanese terrorists, who had machine-gunned passengers picking up their checked luggage at the Lod Airport in Tel Aviv on May 30, 1972.

15. Mad Cow and Other Prion Strains

1. R. G. Will, J. W. Ironside, M. Zeidler, S. N. Cousens, K. Estibeiro, A. Alperovitch, S. Poser, M. Pocchiari, A. Hofman, and P. G. Smith, "A new variant of Creutzfeldt-Jakob disease in the UK," *Lancet* 347: 921–25 (1996).

2. R. M. Anderson, C. A. Donnelly, N. M. Ferguson, M. E. J. Woolhouse, C. J. Watt, H. J. Udy, S. MaWhinney, S. P. Dunstan, T. R. E. Southwood, J. W. Wilesmith, J. B. M. Ryan, L. J. Hoinville, J. E. Hillerton, A. R. Austin, and G. A. H. Wells, "Transmission dynamics and epidemiology of BSE in British cattle," *Nature* 382: 779–88 (1996).

3. G. A. H. Wells, A. C. Scott, C. T. Johnson, R. F. Gunning, R. D. Hancock, M. Jeffrey, M. Dawson, and R. Bradley, "A novel progressive spongiform encephalopathy in cattle," *Vet. Rec.* 121: 419–20 (1987).

4. J. Hope, L. J. D. Reekie, N. Hunter, G. Multhaup, K. Beyreuther, H. White, A. C. Scott, M. J. Stack, M. Dawson, and G. A. H. Wells, "Fibrils from brains of cows with new cattle disease contain scrapie-associated protein," *Nature* 336: 390–92 (1988).

5. The three other members were John Walton, a University of Newcastle neurologist; Anthony Epstein, a retired virologist from Bristol University; and William Martin, a veterinary scientist and director of the Moredun Institute of Animal Diseases in Edinburgh. (Some agriculture officials thought Richard Kimberlin should be a member of the committee, but others saw his dismissal of prions as detrimental; in the end, he was used as a frequent consultant.)

6. J. W. Wilesmith, J. B. M. Ryan, and M. J. Atkinson, "Bovine spongiform encephalopathy: Epidemiologic studies on the origin," *Vet. Rec.* 128: 199–203 (1991).

7. C. J. Gibbs, Jr., D. C. Gajdusek, and R. Latarjet, "Unusual resistance to ionizing radiation of the viruses of kuru, Creutzfeld-Jakob disease, and scrapie," *Proc. Nat. Acad. Sci.* 75: 6268–70 (1978).

8. J. M. Bockman, D. T. Kingsbury, M. P. McKinley, P. E. Bendheim, and S. B. Prusiner, "Creutzfeldt-Jakob disease prion proteins in human brains," *N. Engl. J. Med.* 312: 73–78 (1985).

9. T. K. Koch, B. O. Berg, S. J. DeArmond, and R. F. Gravina, "Creutzfeldt-Jakob disease in a young adult with idiopathic hypopituitarism. Possible relation to the administration of cadaveric human growth hormone," *N. Engl. J. Med.* 313: 731–33 (1985).

10. I. H. Pattison and G. C. Millson, "Scrapie produced experimentally in goats with special reference to the clinical syndrome," *J. Comp. Pathol.* 71: 101–8 (1961).

11. S. B. Prusiner, "Molecular biology of prion diseases," *Science,* New Series 252: 1515–22 (1991).

12. R. F. Marsh, R. A. Bessen, S. Lehmann, and G. R. Hartsough, "Epidemiological and experimental studies on a new incident of transmissible mink encephalopathy," *J. Gen. Virol.* 72: 589–94 (1991).

13. R. A. Bessen and R. F. Marsh, "Identification of two biologically distinct strains of transmissible mink encephalopathy in hamsters," *J. Gen. Virol.* 73: 329–34 (1992).

14. G. C. Telling, P. Parchi, S. J. DeArmond, P. Cortelli, P. Montagna, R. Gabizon, J. Mastrianni, E. Lugaresi, P. Gambetti, and S. B. Prusiner, "Evidence for the conformation of the pathologic isoform of the prion protein enciphering and propagating prion diversity," *Science* 274: 2079–82 (1996).

15. J. A. Mastrianni, R. Nixon, R. Layzer, G. C. Telling, D. Han, S. J. DeArmond, and S. B. Prusiner, "Prion protein conformation in a patient with sporadic fatal insomnia," *N. Engl. J. Med.* 340: 1630–38 (1999).

16. J. G. Safar, S. J. DeArmond, K. Kociuba, C. Deering, S. Didorenko, E. Bouzamondo-Bernstein, S. B. Prusiner, and P. Tremblay, "Prion clearance in bigenic mice," *J. Gen. Virol.* 86: 2913–23 (2005).

17. R. Setlow and B. Doyle, "The action of monochromatic ultraviolet light on proteins," *Biochim. Biophys. Acta* 24: 27–41 (1957); R. B. Setlow, "Shedding light on proteins, nucleic acids, cells, humans, and fish," *Mutat. Res.* 511: 1–14 (2002).

18. D. W. Colby and S. B. Prusiner, "*De novo* generation of prions," *Nat. Rev. Microbiol.* 9: 771–77 (2011).

16. Stockholm

1. In 2013, Tom Südhof received both the Lasker Award and the Nobel Prize; the only other instance appears to have been Marshall Nirenberg in 1968.

2. P. Tremblay, Z. Meiner, M. Galou, C. Heinrich, C. Petromilli, T. Lisse, J. Cayetano, M. Torchia, W. Mobley, H. Bujard, S. J. DeArmond, and S. B. Prusiner, "Doxycyline control of prion protein transgene expression modulates prion disease in mice," *Proc. Natl. Acad. Sci. USA* 95: 12580–85 (1998).

3. J. S. Griffith, "Nature of the scrapie agent: Self-replication and scrapie," *Nature* 215: 1043–44 (1967).

4. R. Rhodes, "Pathological science," *New Yorker,* December 1, 1997, pp. 54–59.

5. G. Taubes, "Nobel gas," *Slate,* October 11, 1997, http://www.slate.com /articles/news_and_politics/hey_wait_a_minute/1997/10/nobel_gas.single .html.

6. J. C. Venter, *A Life Decoded: My Genome: My Life* (Penguin, New York, 2007).

7. D. Fo, "Against jesters who defame and insult (Les Prix Nobel Lecture)," in *Les Prix Nobel,* ed. T. Frängsmyr (Almqvist and Wiksell International, Stockholm, 1997), 381–406.

17. The Third Judgment of Paris

1. M. Jucker and Y. Christen, eds., *Proteopathic Seeds and Neurodegenerative Diseases* (Springer-Verlag, Heidelberg, 2013).

2. S. B. Prusiner, "Scrapie prions," *Annu. Rev. Microbiol.* 43: 345–74 (1989); S. B. Prusiner, "A unifying role for prions in neurodegenerative diseases," *Science* 336: 1511–13 (2012); S. B. Prusiner, "Biology and genetics of prions causing neurodegeneration," *Annu. Rev. Genet.* 47: 601–23 (2013).

3. S. B. Prusiner, "Shattuck Lecture—Neurodegenerative diseases and prions," *N. Engl. J. Med.* 344: 1516–26 (2001).

4. Prusiner, "Biology and genetics of prions causing neurodegeneration."

5. Among the suggestions offered to explain late-onset neurodegeneration: age-dependent mitochondrial DNA mutations; oxidative modifications of DNA, lipids, and/or proteins; proteasome and/or lysosome malfunction; modified innate immunity; exogenous toxins such as alcohol and drugs; concomitant conditions such as atherosclerosis; somatic mutations; RNA-DNA differences; chaperone malfunction; and haploinsufficiency. [Prusiner, "Biology and genetics of prions causing neurodegeneration."]

6. R. W. Mahley, K. H. Weisgraber, and Y. Huang, "Apolipoprotein E4: A causative factor and therapeutic target in neuropathology, including Alzheimer's disease," *Proc. Natl. Acad. Sci. USA* 103: 5644–51 (2006).

7. J. M. Lee, E. M. Ramos, J. H. Lee, T. Gillis, J. S. Mysore, M. R. Hayden, S. C. Warby, P. Morrison, M. Nance, C. A. Ross, R. L. Margolis, F. Squitieri, S. Orobello, S. Di Donato, E. Gomez-Tortosa, C. Ayuso, O. Suchowersky, R. J. Trent, E. McCusker, A. Novelletto, M. Frontali, R. Jones, T. Ashizawa, S. Frank, M. H. Saint-Hilaire, S. M. Hersch, H. D. Rosas, D. Lucente, M. B. Harrison, A. Zanko, R. K. Abramson, K. Marder, J. Sequeiros, J. S. Paulsen, G. B. Landwehr-meyer, R. H. Myers, M. E. Macdonald, and J. F. Gusella, "CAG repeat expansion in Huntington disease determines age at onset in a fully dominant fashion," *Neurology* 78: 690–95 (2012).

8. P. H. Ren, J. E. Lauckner, I. Kachirskaia, J. E. Heuser, R. Melki, and R. R. Kopito, "Cytoplasmic penetration and persistent infection of mammalian cells by polyglutamine aggregates," *Nat. Cell Biol.* 11: 219–25 (2009).

9. H. Braak, E. Braak, D. Yilmazer, R. A. de Vos, E. N. Jansen, and J. Bohl, "Pattern of brain destruction in Parkinson's and Alzheimer's diseases," *J. Neural Transm.* 103: 455–90 (1996).

10. W. W. Seeley, R. K. Crawford, J. Zhou, B. L. Miller, and M. D. Greicius, "Neurodegenerative diseases target large-scale human brain networks," *Neuron* 62: 42–52 (2009).

11. A. de Calignon, M. Polydoro, M. Suarez-Calvet, C. William, D. H. Adamowicz, K. J. Kopeikina, R. Pitstick, N. Sahara, K. H. Ashe, G. A. Carlson, T. L. Spires-Jones, and B. T. Hyman, "Propagation of tau pathology in a model of early Alzheimer's disease," *Neuron* 73: 685–97 (2012); L. Liu, V. Drouet, J. W. Wu, M. P. Witter, S. A. Small, C. Clelland, and K. Duff, "Trans-synaptic spread of tau pathology *in vivo*," *PLoS ONE* 7: e31302 (2012).

12. R. B. Wickner, "[URE3] as an altered URE2 protein: evidence for a prion analog in *Saccharomyces cerevisiae*," *Science* 264: 566–69 (1994).

13. R. M. Ridley, H. F. Baker, C. P. Windle, and R. M. Cummings, "Very long term studies of the seeding of beta-amyloidosis in primates," *J. Neural Transm.* 113: 1243–51 (2006).

14. M. Meyer-Luehmann, J. Coomaraswamy, T. Bolmont, S. Kaeser, C. Schaefer, E. Kilger, A. Neuenschwander, D. Abramowski, P. Frey, A. L. Jaton, J. M. Vigouret, P. Paganetti, D. M. Walsh, P. M. Mathews, J. Ghiso, M. Staufenbiel, L. C. Walker, and M. Jucker, "Exogenous induction of cerebral beta-amyloidogenesis is governed by agent and host," *Science* 313: 1781–84 (2006); Y. S. Eisele, T. Bolmont, M. Heikenwalder, F. Langer, L. H. Jacobson, Z. X. Yan, K. Roth, A. Aguzzi, M. Staufenbiel, L. C. Walker, and M. Jucker, "Induction of cerebral β-amyloidosis: Intracerebral versus systemic Aβ inoculation," *Proc. Natl. Acad. Sci. USA* 106: 12926–31 (2009); R. F. Rosen, J. J. Fritz, J. Dooyema, A. F. Cintron, T. Hamaguchi, J. J. Lah, H. Levine, 3rd, M. Jucker, and L. C. Walker, "Exogenous seeding of cerebral beta-amyloid deposition in betaAPP-transgenic rats," *J. Neurochem.* 120: 660–66 (2012); R. Morales, C. Duran-Aniotz, J. Castilla, L. D. Estrada, and C. Soto, "*De novo* induction of amyloid-β deposition *in vivo*," *Mol. Psychiatry* 1347–53 (2012).

15. G. Tamgüney, K. P. Francis, K. Giles, A. Lemus, S. J. DeArmond, and S. B. Prusiner, "Measuring prions by bioluminescence imaging," *Proc. Natl. Acad. Sci. USA* 106: 15002–6 (2009).

16. J. C. Watts, K. Giles, S. K. Grillo, A. Lemus, S. J. DeArmond, and S. B. Prusiner, "Bioluminescence imaging of Abeta deposition in bigenic mouse models of Alzheimer's disease," *Proc. Natl. Acad. Sci. USA* 108: 2528–33 (2011).

17. D. W. Colby, K. Giles, G. Legname, H. Wille, I. V. Baskakov, S. J. DeArmond, and S. B. Prusiner, "Design and construction of diverse mammalian prior strains," *Proc. Natl. Acad. Sci. USA* 106:20417–22 (2009).

18. J. Stöhr, J. C. Watts, Z. L. Mensinger, A. Oehler, S. K. Grillo, S. J. DeArmond, S. B. Prusiner, and K. Giles, "Purified and synthetic Alzheimer's amyloid beta (Aβ) prions," *Proc. Natl. Acad. Sci. USA* 109: 11025–30 (2012).

19. F. Clavaguera, T. Bolmont, R. A. Crowther, D. Abramowski, S. Frank, A. Probst, G. Fraser, A. K. Stalder, M. Beibel, M. Staufenbiel, M. Jucker,

M. Goedert, and M. Tolnay, "Transmission and spreading of tauopathy in transgenic mouse brain," *Nat. Cell Biol.* 11: 909–13 (2009).

20. B. Frost, R. L. Jacks, and M. I. Diamond, "Propagation of tau misfolding from the outside to the inside of a cell," *J. Biol. Chem.* 284: 12845–52 (2009); J. L. Guo and V. M.-Y. Lee, "Seeding of normal tau by pathological tau conformers drives pathogenesis of Alzheimer-like tangles," *J. Biol. Chem.* 286: 15317–31 (2011).

21. M. Hutton, C. L. Lendon, P. Rizzu, M. Baker, S. Froelich, H. Houlden, S. Pickering-Brown, S. Chakraverty, A. Isaacs, A. Grover, J. Hackett, J. Adamson, S. Lincoln, D. Dickson, P. Davies, R. C. Petersen, M. Stevens, E. de Graaff, E. Wauters, J. van Baren, M. Hillebrand, M. Joosse, J. M. Kwon, P. Nowotny, L. K. Che, J. Norton, J. C. Morris, L. A. Reed, J. Trojanowski, H. Basun, L. Lannfelt, M. Neystat, S. Fahn, F. Dark, T. Tannenberg, P. R. Dodd, N. Hayward, J. B. J. Kwok, P. R. Schofield, A. Andreadis, J. Snowden, D. Craufurd, D. Neary, F. Owen, B. A. Oostra, J. Hardy, A. Goate, J. van Swieten, D. Mann, T. Lynch, and P. Heutink, "Association of missense and 5'-splice-site mutations in tau with the inherited dementia FTDP-17," *Nature* 393: 702–5 (1998).

22. K. Rascovsky, J. R. Hodges, D. Knopman, M. F. Mendez, J. H. Kramer, J. Neuhaus, J. C. van Swieten, H. Seelaar, E. G. Dopper, C. U. Onyike, A. E. Hillis, K. A. Josephs, B. F. Boeve, A. Kertesz, W. W. Seeley, K. P. Rankin, J. K. Johnson, M. L. Gorno-Tempini, H. Rosen, C. E. Prioleau-Latham, A. Lee, C. M. Kipps, P. Lillo, O. Piguet, J. D. Rohrer, M. N. Rossor, J. D. Warren, N. C. Fox, D. Galasko, D. P. Salmon, S. E. Black, M. Mesulam, S. Weintraub, B. C. Dickerson, J. Diehl-Schmid, F. Pasquier, V. Deramecourt, F. Lebert, Y. Pijnenburg, T. W. Chow, F. Manes, J. Grafman, S. F. Cappa, M. Freedman, M. Grossman, and B. L. Miller, "Sensitivity of revised diagnostic criteria for the behavioural variant of frontotemporal dementia," *Brain* 134: 2456–77 (2011).

23. B. I. Omalu, S. T. DeKosky, R. L. Minster, M. I. Kamboh, R. L. Hamilton, and C. H. Wecht, "Chronic traumatic encephalopathy in a National Football League player," *Neurosurgery* 57: 128–34 (2005); L. N. Hazrati, M. C. Tartaglia, P. Diamandis, K. D. Davis, R. E. Green, R. Wennberg, J. C. Wong, L. Ezerins, and C. H. Tator, "Absence of chronic traumatic encephalopathy in retired football players with multiple concussions and neurological symptomatology," *Front. Hum. Neurosci.* 7: 222 (2013).

24. A. C. McKee, R. C. Cantu, C. J. Nowinski, E. T. Hedley-Whyte, B. E. Gavett, A. E. Budson, V. E. Santini, H. S. Lee, C. A. Kubilus, and R. A. Stern, "Chronic traumatic encephalopathy in athletes: Progressive tauopathy after repetitive head injury," *J. Neuropathol. Exp. Neurol.* 68: 709–35 (2009).

25. B. Omalu, J. L. Hammers, J. Bailes, R. L. Hamilton, M. I. Kamboh, G. Webster, and R. P. Fitzsimmons, "Chronic traumatic encephalopathy in an Iraqi war veteran with posttraumatic stress disorder who committed suicide," *Neurosurg. Focus* 31: E3 (2011).

26. A. C. McKee, T. D. Stein, C. J. Nowinski, R. A. Stern, D. H. Daneshvar, V. E. Alvarez, H. S. Lee, G. Hall, S. M. Wojtowicz, C. M. Baugh, D. O. Riley, C. A. Kubilus, K. A. Cormier, M. A. Jacobs, B. R. Martin, C. R. Abraham, T. Ikezu, R. R. Reichard, B. L. Wolozin, A. E. Budson, L. E. Goldstein, N. W. Kowall, and R. C. Cantu, "The spectrum of disease in chronic traumatic encephalopathy," *Brain* 136: 43–64 (2013).

27. C. W. Hoge, D. McGurk, J. L. Thomas, A. L. Cox, C. C. Engel, and C. A. Castro, "Mild traumatic brain injury in U.S. soldiers returning from Iraq," *N. Engl. J. Med.* 358: 453–63 (2008).

28. M. Haiken, "Suicide rate among vets and active duty military jumps—now 22 a day," *Forbes,* February 5, 2013.

29. H. Braak and E. Braak, "Staging of Alzheimer's disease-related neurofibrillary changes," *Neurobiol. Aging* 16: 271–84 (1995); M. P. Lambert, A. K. Barlow, B. A. Chromy, C. Edwards, R. Freed, M. Liosatos, T. E. Morgan, I. Rozovsky, B. Trommer, K. L. Viola, P. Wals, C. Zhang, C. E. Finch, G. A. Krafft, and W. L. Klein, "Diffusible, nonfibrillar ligands derived from A 1–42 are potent central nervous system neurotoxins," *Proc. Natl. Acad. Sci. USA* 95: 6448–53 (1998); M. Rapoport, H. N. Dawson, L. I. Binder, M. P. Vitek, and A. Ferreira, "Tau is essential to beta-amyloid-induced neurotoxicity," *Proc. Natl. Acad. Sci. USA* 99: 6364–69 (2002).

30. K. Hsiao, S. Dlouhy, M. R. Farlow, C. Cass, M. Da Costa, M. Conneally, M. E. Hodes, B. Ghetti, and S. B. Prusiner, "Mutant prion proteins in Gerstmann-Sträussler-Scheinker disease with neurofibrillary tangles," *Nat. Genet.* 1: 68–71 (1992); S. Mead, S. Gandhi, J. Beck, D. Caine, D. Gallujipali, C. Carswell, H. Hyare, S. Joiner, H. Ayling, T. Lashley, J. M. Linehan, H. Al-Doujaily, B. Sharps, T. Revesz, M. K. Sandberg, M. M. Reilly, M. Koltzenburg, A. Forbes, P. Rudge, S. Brandner, J. D. Warren, J. D. Wadsworth, N. W. Wood, J. L. Holton, and J. Collinge, "A novel prion disease associated with diarrhea and autonomic neuropathy," *N. Engl. J. Med.* 369: 1904–14 (2013).

31. Lewy bodies were first described in 1912 by neurologist Frederic Lewy after he examined the brains of twenty-five patients who had died of Parkinson's disease. At the time, Lewy was director of the Neuropsychiatric Laboratory at the Medical School of Breslau University under Alois Alzheimer. In 1932 Lewy was appointed director of the Neurological Research Institute and Clinic in Berlin, but he was dismissed by the Nazis on July 1,

1933. After a short stay in England, he immigrated to the United States and obtained a job in the Department of Neurosurgery at the University of Pennsylvania, where he worked until he died in 1950 at the age of sixty-five; B. Holdorff, "Friedrich Heinrich Lewy (1885–1950) and his work," *J. Hist. Neurosci.* 11: 19–28 (2002).

32. J. Y. Li, E. Englund, J. L. Holton, D. Soulet, P. Hagell, A. J. Lees, T. Lashley, N. P. Quinn, S. Rehncrona, A. Bjorklund, H. Widner, T. Revesz, O. Lindvall, and P. Brundin, "Lewy bodies in grafted neurons in subjects with Parkinson's disease suggest host-to-graft disease propagation," *Nat. Med.* 14: 501–3 (2008).

33. J. H. Kordower, Y. Chu, R. A. Hauser, T. B. Freeman, and C. W. Olanow, "Lewy body-like pathology in long-term embryonic nigral transplants in Parkinson's disease," *Nat. Med.* 14: 504–6 (2008).

34. M. H. Polymeropoulos, C. Lavedan, E. Leroy, S. E. Ide, A. Dehejia, A. Dutra, B. Pike, H. Root, J. Rubenstein, R. Boyer, E. S. Stenroos, S. Chandrasekharappa, A. Athanassiadou, T. Papapetropoulos, W. G. Johnson, A. M. Lazzarini, R. C. Duvoisin, G. Di Iorio, L. I. Golbe, and R. L. Nussbaum, "Mutation in the α-synuclein gene identified in families with Parkinson's disease," *Science* 276: 2045–47 (1997); M. G. Spillantini, M. L. Schmidt, V. M.-Y. Lee, J. Q. Trojanowski, R. Jakes, and M. Goedert, "α-Synuclein in Lewy bodies," *Nature* 388: 839–40 (1997).

35. C. W. Olanow and S. B. Prusiner, "Is Parkinson's disease a prion disorder?" *Proc. Natl. Acad. Sci. USA* 106: 12571–72 (2009).

36. H. Braak, K. Del Tredici, U. Rub, R. A. de Vos, E. N. Jansen Steur, and E. Braak, "Staging of brain pathology related to sporadic Parkinson's disease," *Neurobiol. Aging* 24: 197–211 (2003).

37. A.-L. Mougenot, S. Nicot, A. Bencsik, E. Morignat, J. Verchère, L. Lakhdar, S. Legastelois, and T. Baron, "Prion-like acceleration of a synucleinopathy in a transgenic mouse model," *Neurobiol. Aging* 33: 2225–28 (2012).

38. L. A. Volpicelli-Daley, K. C. Luk, T. P. Patel, S. A. Tanik, D. M. Riddle, A. Stieber, D. F. Meaney, J. Q. Trojanowski, and V. M. Lee, "Exogenous alpha-synuclein fibrils induce Lewy body pathology leading to synaptic dysfunction and neuron death," *Neuron* 72: 57–71 (2011); K. C. Luk, V. M. Kehm, B. Zhang, P. O'Brien, J. Q. Trojanowski, and V. M. Y. Lee, "Intracerebral inoculation of pathological α-synuclein initiates a rapidly progressive neurodegenerative α-synucleinopathy in mice," *J. Exp. Med.* 209: 975–86 (2012).

39. K. C. Luk, V. Kehm, J. Carroll, B. Zhang, P. O'Brien, J. Q. Trojanowski, and V. M. Lee, "Pathological alpha-synuclein transmission initiates Parkinson-like neurodegeneration in nontransgenic mice," *Science* 338: 949–53 (2012); M. Masuda-Suzukake, T. Nonaka, M. Hosokawa, T. Oikawa, T. Arai, H. Akiyama,

D. M. Mann, and M. Hasegawa, "Prion-like spreading of pathological alpha-synuclein in brain," *Brain* 136: 1128–38 (2013).

40. J. C. Watts, K. Giles, A. Oehler, L. T. Middleton, D. T. Dexter, S. M. Gentleman, S. J. DeArmond, and S. B. Prusiner, "Transmission of multiple system atrophy prions to transgenic mice," *Proc. Natl. Acad. Sci. USA.* 110: 19555–60 (2013).

41. C. Münch, J. O'Brien, and A. Bertolotti, "Prion-like propagation of mutant superoxide dismutase-1 misfolding in neuronal cells," *Proc. Natl. Acad. Sci. USA* 108: 3548–53 (2011); L. I. Grad, W. C. Guest, A. Yanai, E. Pokrishevsky, M. A. O'Neill, E. Gibbs, V. Semenchenko, M. Yousefi, D. S. Wishart, S. S. Plotkin, and N. R. Cashman, "Intermolecular transmission of superoxide dismutase 1 misfolding in living cells," *Proc. Natl. Acad. Sci. USA* 108: 16398–403 (2011).

42. K. Si, Y. B. Choi, E. White-Grindley, A. Majumdar, and E. R. Kandel, "Aplysia CPEB can form prion-like multimers in sensory neurons that contribute to long-term facilitation," *Cell* 140: 421–35 (2010).

43. N. Gilks, N. Kedersha, M. Ayodele, L. Shen, G. Stoecklin, L. M. Dember, and P. Anderson, "Stress granule assembly is mediated by prion-like aggregation of TIA-1," *Mol. Biol. Cell.* 15: 5383–98 (2004); F. Hou, L. Sun, H. Zheng, B. Skaug, Q. X. Jiang, and Z. J. Chen, "MAVS forms functional prion-like aggregates to activate and propagate antiviral innate immune response," *Cell* 146: 448–61 (2011).

44. M. Jucker and L. C. Walker, "Self-propagation of pathogenic protein aggregates in neurodegenerative diseases," *Nature* 501: 45–51 (2013); Prusiner, "Biology and genetics of prions causing neurodegeneration."

45. S. B. Prusiner, M. Scott, D. Foster, K.-M. Pan, D. Groth, C. Mirenda, M. Torchia, S.-L. Yang, D. Serban, G. A. Carlson, P. C. Hoppe, D. Westaway, and S. J. DeArmond, "Transgenetic studies implicate interactions between homologous PrP isoforms in scrapie prion replication," *Cell* 63: 673–86 (1990).

46. L. F. Lue, Y. M. Kuo, A. E. Roher, L. Brachova, Y. Shen, L. Sue, T. Beach, J. H. Kurth, R. E. Rydel, and J. Rogers, "Soluble amyloid beta peptide concentration as a predictor of synaptic change in Alzheimer's disease," *Am. J. Pathol.* 155: 853–62 (1999); C. A. McLean, R. A. Cherny, F. W. Fraser, S. J. Fuller, M. J. Smith, K. Beyreuther, A. I. Bush, and C. L. Masters, "Soluble pool of Abeta amyloid as a determinant of severity of neurodegeneration in Alzheimer's disease," *Ann. Neurol.* 46: 860–66 (1999).

47. W. Zhang, J. Hao, R. Liu, Z. Zhang, G. Lei, C. Su, J. Miao, and Z. Li, "Soluble Abeta levels correlate with cognitive deficits in the 12-month-old APPswe/PS1dE9 mouse model of Alzheimer's disease," *Behav. Brain Res.* 222: 342–50 (2011); Stöhr et al. "Purified and synthetic Alzheimer's amyloid beta (Aβ) prions."

48. S. Nath, L. Agholme, F. R. Kurudenkandy, B. Granseth, J. Marcusson, and M. Hallbeck, "Spreading of neurodegenerative pathology via neuron-to-neuron transmission of beta-amyloid," *J. Neurosci.* 32: 8767–77 (2012).

49. Y. Kawasaki, K. Kawagoe, C. J. Chen, K. Teruya, Y. Sakasegawa, and K. Doh-ura, "Orally administered amyloidophilic compound is effective in prolonging the incubation periods of animals cerebrally infected with prion diseases in a prion strain-dependent manner," *J. Virol.* 81: 12889–98 (2007); D. Lu, K. Giles, Z. Li, S. Rao, E. Dolghih, J. R. Gever, M. Geva, M. Elepano, A. Oehler, C. Bryant, A. R. Renslo, M. P. Jacobson, S. J. DeArmond, B. M. Silber, and S. B. Prusiner, "Biaryl amides and hydrazones as therapeutics for prion disease in transgenic mice," *J. Pharmcol. Exp. Therap.* 347: 325–38 (2013); D. B. Berry, D. Lu, M. Geva, J. C. Watts, S. Bhardwaj, A. Oehler, A. R. Renslo, S. J. DeArmond, S. B. Prusiner, and K. Giles, "Drug resistance confounding prion therapeutics," *Proc. Natl. Acad. Sci. USA.* 110: E4160–69 (2013).

50. S. B. Prusiner, M. P. McKinley, K. A. Bowman, D. C. Bolton, P. E. Bendheim, D. F. Groth, and G. G. Glenner, "Scrapie prions aggregate to form amyloid-like birefringent rods," *Cell* 35: 349–58 (1983); S. B. Prusiner, "Some speculations about prions, amyloid, and Alzheimer's disease," *N. Engl. J. Med.* 310: 661–63 (1984).

51. Prusiner, "Shattuck Lecture—Neurodegenerative diseases and prions"; S. B. Prusiner, "A unifying role for prions in neurodegenerative diseases," *Science* 336: 1511–13 (2012); Prusiner, "Biology and genetics of prions causing neurodegeneration."

52. I. Berlin, *The Hedgehog and the Fox: An Essay on Tolstoy's View of History* (Weidenfeld and Nicolson, London, 1953).

Epilogue

1. T. C. Chamberlin, "The method of multiple working hypotheses," *Science [Old Series]* 15: 92–97 (1890).

2. M. D. Hurd, P. Martorell, A. Delavande, K. J. Mullen, and K. M. Langa, "Monetary costs of dementia in the United States," *N. Engl. J. Med.* 368: 1326–34 (2013).

3. M. Wachterman, D. K. Kiely, and S. L. Mitchell, "Reporting dementia on the death certificates of nursing home residents dying with end-stage dementia," *JAMA* 300: 2608–10 (2008).

4. B. D. James, S. E. Leurgans, L. E. Hebert, P. A. Scherr, K. Yaffe, and D. A. Bennett, "The contribution of Alzheimer's disease to mortality in the United States," *Neurology* (in press).

5. S. D. O'Connor, S. B. Prusiner, and K. Dychtwald, "The age of Alzheimer's," *New York Times,* October 28, 2010.

6. R. D. Terry, "Dementia," *Arch. Neurol.* 33: 1–4 (1976); R. Katzman, "Editorial: The prevalence and malignancy of Alzheimer disease. A major killer," *Arch. Neurol.* 33: 217–28 (1976); R. Katzman, "Medical progress: Alzheimer's disease," *N. Engl. J. Med.* 314: 964–73 (1986).

7. A. Alzheimer, "Ueber eine eigenartige Erkrankung der Hirnrinde," *Cent. Nervenheilk. Psychiat.* 30: 177–79 (1907); G. G. Glenner and C. W. Wong, "Alzheimer's disease: Initial report of the purification and characterization of a novel cerebrovascular amyloid protein," *Biochem. Biophys. Res. Commun.* 120: 885–90 (1984); J.-P. Brion, H. Passareiro, J. Nunez, and J. Flament-Durand, "Mise en évidence immunologique de la protéine tau au niveau des lésions de dégénérescence neurofibrillaire de la maladie d'Alzheimer," *Arch. Biol.* 95: 229–35 (1985); K. S. Kosik, C. L. Joachim, and D. J. Selkoe, "Microtubule-associated protein tau is a major antigenic component of paired helical filaments in Alzheimer disease," *Proc. Natl. Acad. Sci. USA* 83: 4044–48 (1986); I. Grundke-Iqbal, K. Iqbal, Y.-C. Tung, M. Quinlan, H. M. Wisniewski, and L. I. Binder, "Abnormal phosphorylation of the microtubule-associated protein (tau) in Alzheimer cytoskeletal pathology," *Proc. Natl. Acad. Sci. USA* 83: 4913–17 (1986); J. G. Wood, S. S. Mirra, N. J. Pollock, and I. I. Binder, "Neurofibrillary tangles of Alzheimer's disease share antigenic determinants with the axonal microtubule-associated protein tau," *Proc. Natl. Acad. Sci. USA* 83: 4040–43 (1986).

8. S. B. Prusiner and G. P. Shultz, "Brain research and the challenge of aging," *Wall Street Journal*, July 5, 2011.

9. C. W. Olanow and S. B. Prusiner, "Is Parkinson's disease a prion disorder?" *Proc. Natl. Acad. Sci. USA* 106: 12571–72 (2009).

10. D. A. Evans, H. H. Funkenstein, M. S. Albert, P. A. Scherr, N. R. Cook, M. J. Chown, L. E. Hebert, C. H. Hennekens, and J. O. Taylor, "Prevalence of Alzheimer's disease in a community population of older persons—Higher than previously reported," *JAMA* 10: 2551–56 (1989).

11. C. W. Hoge, D. McGurk, J. L. Thomas, A. L. Cox, C. C. Engel, and C. A. Castro, "Mild traumatic brain injury in U.S. soldiers returning from Iraq," *N. Engl. J. Med.* 358: 453–63 (2008).

12. M. Jucker and L. C. Walker, "Self-propagation of pathogenic protein aggregates in neurodegenerative diseases," *Nature* 501: 45–51 (2013); S. B. Prusiner, "Biology and genetics of prions causing neurodegeneration," *Annu. Rev. Genet.* 47: 601–23 (2013)

Glossary

allele: one of two or more forms of a gene.

alpha helix: structural element of a protein, in which the polypeptide backbone follows a right-handed helical path approximating a cylinder, with a periodicity of 3.6 amino acids per turn. In this conformation of the polypeptide chain, successive turns are held together by hydrogen bonds.

Alzheimer's disease (AD): most common, progressive neurodegenerative disorder characterized by impaired recent memory, confusion, visual-spatial disorientation, language dysfunction progressing from anomia to fluent aphasia, difficulties with calculations, and deterioration of judgment. Later, impaired long-term memory, delusions, and hallucinations may occur. AD is the most common degenerative brain disorder accounting for more than 70 percent of all cases of dementia. Onset is typically in the seventies and eighties, and death generally occurs in five to ten years from diagnosis. Neuropathologically, amyloid plaques composed largely of Aβ fibrils and neurofibrillary tangles containing paired helical filaments comprising tau prions.

amino acid: organic acid containing one or more amino substituents. Twenty different, naturally occurring amino acids are found in proteins.

amino acid sequence: assembly of amino acids into a chain called a protein. The sequence is specified by the order of bases in messenger RNA according to the genetic code. The sequence of a protein is also called the primary structure.

amyloid: proteinaceous fibrils, identified by their tinctorial, ultrastructural, and structural properties. Amyloid deposition is a feature of most neurodegenerative diseases; the amyloid of each disease is composed of a different protein. Some of the proteins that form amyloids can turn into prions and cause neurodegeneration.

amyloid beta (Aβ): Aβ peptide is composed of approximately forty amino acids. This peptide is cleaved from a much longer protein called the amyloid precursor protein (APP) by the β- and γ-secretases. Aβ can form prions and cause Alzheimer's disease.

amyloid plaque: collection of amyloid fibrils in the extracellular space of the brain. In Alzheimer's disease such plaques are composed of Aβ fibrils and in Creutzfeldt-Jakob disease of PrP^{Sc} fibrils.

antibody: immunoglobulin (Ig), which is a large Y-shaped protein produced by B-lymphocytes that is used by the immune system to identify and neutralize foreign substances such as bacteria and viruses. The antibody recognizes a unique part of the foreign target, called an antigen. Scientists often use the specificity of antibodies to detect and quantify individual proteins.

argyrophilic grain disease: late-onset neurodegeneration caused by tau prions. Shares clinical features of both Alzheimer's and frontotemporal dementias: reduction in short-term memory; difficulty finding words, reading, and writing; disorientation; behavioral disturbances including personality changes, emotional disorders with aggression, and ill-temper. Silver-stained (argyrophilic) grains or coiled bodies within the cytoplasm of neurons consisting mainly of tau are found in the amygdala, hippocampus, and entorhinal cortex.

ataxia: neurological sign consisting of lack of voluntary coordination of muscle movements, as in walking. Ataxia is often due to lesions in the cerebellum, which coordinates movements. Lesions in the center or vermis of the cerebellum result in truncal ataxia (also called titubation) as well as tremor of the head, while lesions in hemispheres produce limb ataxia.

autosomal: pattern of inheritance for a genetic trait encoded in a gene located on any chromosome other than the X and Y sex chromosomes.

base pair: two compounds on opposite complementary DNA or certain types of RNA strands that are connected via hydrogen bonds are called a base pair (bp). In DNA base pairing, adenine (A) forms a base pair with thymine (T), and guanine (G) forms a base pair with cytosine (C). In RNA, thymine is replaced by uracil (U).

beta sheet: structural element of a protein in which adjacent β-strands lie approximately flat in either a parallel or antiparallel orientation to each other. Beta sheet regions of proteins are extended and stabilized by hydrogen bonding between NH and CO groups of different polypeptide chain backbones or separate regions of the same chain.

bioluminescence imaging: technique used to image organs wherein luciferase is produced in response to a specific stimulus. Luciferase catalyzes the oxidation of luciferin, resulting in the emission of visible light.

bovine spongiform encephalopathy (BSE): prion disease in cattle caused by the accumulation of bovine PrP^{Sc}. BSE was first reported in 1986 in Great Britain; characterized clinically by apprehensive behavior, hyperesthesia, and ataxia, and histologically by spongiform changes in the gray matter of the brain stem. As in scrapie of sheep and Creutzfeldt-Jakob disease of humans, PrP^{Sc} accumulates in the brain and causes CNS dysfunction. Cases of BSE have been reported in Austria, Britain, Canada, the Czech Republic, Denmark, France, Germany, Ireland, Israel, Italy, Japan, Poland, Portugal, Spain, Switzerland, and the United States.

cerebellum: consists of two lateral hemispheres containing billions of neurons, united by a narrow middle portion, the vermis. It is slightly smaller than a fist and tucked away in the posterior fossa, just above the nape of the neck.

chemical compound: a substance consisting of two or more elements and having a unique structure consisting of a fixed ratio of atoms. Compounds are held together in a defined spatial arrangement by chemical bonds.

chorea: abnormal involuntary movement disorder. Chorea is characterized by brief, irregular movements that are neither repetitive nor rhythmic. Choreiform movements appear to flow from one muscle to the next. The "dance-like" movements of chorea often occur with athetosis, which adds twisting and writhing movements. When chorea and athetosis occur together, they are called choreoathetosis.

chronic traumatic encephalopathy (CTE): progressive neurodegenerative disease that occurs following traumatic brain injuries. CTE is a posttraumatic form of frontotemporal dementia in which tau proteins are converted into prions. Tau prions polymerize into filaments that condense into neurofibrillary tangles.

chronic wasting disease (CWD): neurodegeneration caused by PrP^{Sc} prions in deer, elk, and moose. In contrast to most prion diseases, CWD is highly communicable among cervids.

clone: colony of cells derived from a single cell by asexual reproduction, all having identical genetic constitutions. Homogeneous population of DNA molecules that were copied using recombinant DNA technology.

codon: set of three consecutive nucleotides in a strand of DNA or RNA that provides the genetic information encoding a specific amino acid.

concussion: traumatic brain injury (TBI) that alters brain function. Most concussions are mild TBIs resulting in headache, difficulty concentrating, and impaired memory, judgment, balance, and coordination. Some concussions are accompanied by a loss of consciousness.

conformation: shape of a protein. Patterns of hydrogen bonds determine the secondary structures—i.e., alpha helices and beta sheets—of proteins. The tertiary structure or conformation of a protein develops when the protein folds into an active shape. The driving force for protein folding is the burial of hydrophobic residues in order to prevent them from coming into contact with water.

corticobasal degeneration (CBD): neurodegeneration caused by tau prions. Clinically, patients show Parkinsonism, alien hand syndrome, apraxia as demonstrated by an inability to mimic movements, and aphasia as evidenced by difficulty initiating spoken words. Neuropathologically, inclusions of tau within astrocytes are seen.

Creutzfeldt-Jakob disease (CJD): prion disease in humans caused by the accumulation of human PrP^{Sc}. CJD can be manifest as a genetic (familial CJD), infectious (iatrogenic CJD and variant CJD), or spontaneous (sporadic CJD) disorder. Approximately 85 percent of CJD cases occur sporadically, 10–15 percent are inherited, and less than 1 percent are infectious. The peak incidence is between fifty-five and sixty-five years of age, and the disease is rare before age thirty. Patients with CJD generally present with a progressive dementia as well as multifocal myoclonus. The most effective laboratory tests for antemortem diagnosis use the EEG that demonstrates triphasic sharp waves late in the clinical course and diffusion weighted magnetic resonance imaging that shows hyperintensities in the gray matter. Average survival time is less than one year after onset of symptoms.

dementia: loss of cognitive and intellectual functions without impaired perception or consciousness. Most dementias are progressive and caused by Aβ prions that produce Alzheimer's disease.

denaturation: process by which the native state of a protein or nucleic acid is altered generally resulting in a reduction in biological activity. When heat and/or chemicals unfold enzymes, their catalytic activity diminishes; this unfolding process is called denaturation.

dentatorubral-pallidoluysian atrophy (DRPLA): autosomal dominant progressive dementia with involuntary choreiform movements and ataxia. A polyglutamine expansion in the atrophin-1 (ATN1) protein causes DRPLA, and the longer the expansion, the earlier the onset of disease. At autopsy, the brain is shrunken and microscopic pathology shows intranuclear inclusions composed of mutant ATN1. The excessively long name of this disorder comes from the neurodegenerative changes found in the dentate gyrus, red nucleus, globus pallidus, and the subthalamic nucleus, also called the "body of Luys."

diethylpyrocarbonate (DEPC): chemical reagent that reacts with proteins or nucleic acids. DEPC donates an ethoxyformyl group that results in the inactivation of the protein or nucleic acid. Hydroxylamine removes the ethoxyformyl group from proteins but does not do so with nucleic acids.

DNA (deoxyribonucleic acid): substance of which genes are made. DNA is nucleic acid consisting of deoxyribose molecules esterified with phosphate groups between the 3'- and 5'-hydroxyl groups; linked to this structure are the purine nucleotides adenosine (A) and guanosine (G) and the pyrimidine nucleotides cytosine (C) and thymidine (T). DNA may be open ended or circular, single or double stranded.

dominant disorder: only one mutated copy of the gene is necessary for a person to be afflicted by an autosomal dominant disease. Each afflicted person usually has one affected parent. The chance a child will inherit the mutated gene is 1 in 2 (or 50 percent). When both copies of a gene must be mutated for a person to manifest disease, the disorder is called recessive.

electron microscope: instrument in which electron beams with wavelengths thousands of times shorter than visible light are used to allow much greater resolution and magnification; in this technique the electrons are transmitted through a thin section of an embedded specimen maintained in a vacuum.

electrophoresis: movement of particles in an electric field toward an electric pole (anode or cathode); used to separate and purify biomolecules.

endpoint titration: method for determining infectivity of a sample by bioassay with a serially diluted inoculum. The endpoint is defined as the dilution at which the inoculum is no longer infectious.

enzyme: protein that acts as a catalyst to induce chemical changes in some substance while it itself remains unchanged by the process. Enzymes are composed of one or more chains of amino acids and are active when folded into a particular conformation.

epitope: site on a large molecule to which a specific antibody binds. In proteins, most epitopes are composed of six to eight amino acids.

etiology: cause of disease.

familial Creutzfeldt-Jakob (fCJD): autosomal dominant human prion disease caused by a mutation in the *PRNP* gene that results in the substitution of an amino acid or the insertion of additional amino acid octarepeats. One example of fCJD is that caused by a mutation resulting in the substitution of lysine for glutamic acid at codon 200.

fatal insomnia: human prion disease due to selective degeneration of the dorsal medial nuclei of the thalamus, resulting in reduced sleep. Both sporadic and genetic forms of fatal insomnia have been described. The inherited form of fatal insomnia, referred to as fatal familial insomnia, is caused by a mutation in *PRNP* that results in the substitution of asparagine for aspartic acid at position 178. A methionine at the polymorphic position 129 is also required for the fatal insomnia phenotype. A valine at position 129 prevents the fatal insomnia phenotype but appears to promote neurodegeneration with clinical signs typical of fCJD.

frontotemporal dementias (FTDs): group of progressive, neurodegenerative diseases that affect the frontal and temporal lobes of the brain, gradually destroying the ability to behave appropriately, empathize with others, learn, reason, make judgments, communicate, and plan activities. In people under age sixty, FTD is the second most frequent cause of dementia, with Alzheimer's disease being the most common. Men are affected more commonly than women. Pathologically, the FTDs are subdivided based on the protein deposits identified by immunostaining for tau, TDP43, and FUS. It is likely that all three of these proteins become prions. In FTD caused by tau prions, the subtypes include Pick's disease, corticobasal degeneration (CBD), progressive supranuclear palsy (PSP), frontotemporal dementia and parkinsonism linked to tau mutations (FTDP-17), sporadic multisystem tauopathy, and argyrophilic grain disease.

gene probe: piece of DNA used for locating a particular gene on a chromosome as well as for measuring the expression of the gene.

genetic-linkage study: assessment of the relationship between two loci on the same chromosome to determine whether they are sufficiently close that the respective alleles are not inherited independently by the offspring.

genome: complete set of chromosomes derived from one parent.

germline mutation: alteration in DNA that occurs before conception. Germline mutations occur in the sperm or egg and therefore are passed on to children.

Gerstmann-Sträussler-Scheinker (GSS) disease: autosomal dominant human prion disease caused by a point mutation in *PRNP* that results in the accumulation of PrP^{Sc}. Most if not all GSS patients have PrP amyloid plaques. Several different point mutations causing GSS have been identified, including the substitution of leucine for proline at position 102, valine for alanine at position 117, and serine for phenylalanine at position 198. Although the clinical characteristics seem frequently to be determined by the particular mutation, there are many exceptions. Generally, GSS has a longer, more slowly progressive clinical course compared with CJD, often lasting several years.

glycolipid: fat-soluble substance with one or more covalently attached sugars.

glycoprotein: protein with one or more covalently attached sugars. Generally, complex chains of sugars are attached to glycoproteins.

haploinsufficiency: occurs when there is only a single functional copy of a gene and the other copy is inactivated by mutation. The single functional copy of the gene does not produce enough of a protein to maintain a normal condition; this in turn leads to disease. It is responsible for some but not all autosomal dominant disorders.

HET-s: protein involved in programmed cell death termed heterokaryon incompatibility in the filamentous fungus *Podospora anserina*. For cell death, the HET-s protein undergoes conversion into the prion [Het-s].

homogenate: tissue ground into a creamy consistency in which the cell structure is disrupted. Typically, a 10 percent homogenate is prepared by adding nine parts of a saline solution to one part of cells. The homogenate is a suspension of many different cellular components.

human growth hormone: protein comprising 191 amino acids secreted by the anterior pituitary, which stimulates growth. Human growth hormone is abbreviated HGH; however, in current use HGH refers to preparations prepared using recombinant DNA technology.

huntingtin (Htt): protein comprising 3,144 amino acids containing fewer than thirty-six glutamine residues in the N-terminal region. Expansion of the polyglutamine tract causes Huntington's disease.

Huntington's disease (HD): autosomal-dominant neurodegenerative disease caused by mutant huntingtin prions. Huntington's patients show involuntary writhing movements called chorea as well as progressive dementia and behavioral problems. Patients often manifest anxiety, depression, and diminished executive function, which includes planning and initiating appropriate actions. HD typically becomes noticeable in midadult life but can be manifest in childhood when the polyglutamine-repeat region is very large. Through genetic anticipation, the disease may develop earlier in life in each successive generation. The worldwide prevalence is 5 to 10 patients per 100,000. Neuropathologically, atrophy of the caudate and putamen are seen, as well as nuclear inclusions containing the huntingtin protein.

hydroxylamine: partially oxidized derivative of ammonia that reacts with carbonyl groups to produce oximes. It is a chemical mutagen that causes deamination of cytosine residues in DNA.

iatrogenic CJD (iCJD): form of CJD that is transmitted accidentally through medical procedures, such as the transplantation of prion-contaminated dura mater or other tissue, injection of prion-contaminated human growth hormone from pituitary extracts, or the use of improperly decontaminated surgical instruments or devices.

incubation time: interval between exposure to inoculation of a pathogen and the detection of disease. Generally disease is manifested by clinical symptoms or signs, but noninvasive techniques such as bioluminescence imaging are also used.

infectious unit: smallest number of any pathogenic particles capable of initiating infection in a cell or organism.

intention tremor: increases as an extremity approaches a point or object in a visually guided movement. An intention tremor is usually perpendicular to the direction of movement. Intention tremor is the result of dysfunction of the cerebellum, particularly on the same side as the tremor in the lateral hemispheres, which control visually guided movements. Also called cerebellar tremor.

interferon (IFN): glycoproteins made by host cells in response to pathogens such as viruses and bacteria. IFNs trigger the protective defenses of the immune system by binding to specific receptors on cell membranes and as such are cytokines.

knockout mouse: genetically engineered mouse in which an existing gene has been inactivated, or "knocked out," by replacing it or disrupting it with an artificial piece of DNA. A similar process has been used to render rats deficient for a particular gene.

kuru: progressive, fatal prion disease characterized by ataxia, tremors, lack of coordination, and death. Pathologic lesions in the brain include neuronal loss, astrocytosis, spongiform change, and numerous PrP amyloid plaques primarily in the cerebellum. PrP^{Sc} prions cause kuru, which was confined the Fore and neighboring tribes in the highlands of New Guinea. Kuru prions were transmitted orally through ritualistic cannibalism.

lectin: sugar-binding proteins that are highly specific for particular sugar moieties. In contrast to lectins, glycoproteins contain specific sugars. Lectins are useful tools in the purification of glycoproteins.

Lewy bodies: inclusions in neurons containing α-synuclein assembled into fibrils. Lewy bodies are the pathological hallmark of Parkinson's disease. Mutations in the α-synuclein gene (*SCNA*) cause familial Parkinson's disease. Duplication or triplication of *SCNA* increases α-synuclein levels leading to Parkinson's disease.

Lewy body dementia (LBD): caused by α-synuclein prions and manifest by fluctuating cognition with great variations in attention and alertness, recurrent visual hallucinations, and motor features of Parkinson's. Symptoms of a rapid eye movement (REM)-sleep behavioral disorder are common. Besides Lewy bodies in the dopamine neurons, they are found throughout the cerebral cortex. Brain atrophy occurs as the cerebral cortex degenerates.

light microscope: optical microscope that uses visible light and a lens system to magnify very small objects.

lipid: broad group of naturally occurring molecules that includes fats, waxes and sterols. Lipids may be defined as hydrophobic or amphiphilic small molecules; the amphiphilic nature of some lipids allows them to form such structures as vesicles, liposomes, or membranes in an aqueous environment.

macromolecule: large molecule created by polymerization of smaller subunits. Nucleic acids and proteins are macromolecules formed from chains of nucleotides and amino acids, respectively. The individual constituent molecules of macromolecules are called monomers; both individual nucleotides and amino acids are monomers.

mad cow disease: common name for bovine spongiform encephalopathy.

mass spectrometry (MS): analytical technique that measures the mass-to-charge ratio of charged particles. It is used for determining the masses of particles and elemental composition; it is also used for elucidating the chemical structures of molecules such as peptides. MS works by ionizing chemical compounds to generate charged molecules or molecule fragments and measuring their mass-to-charge ratios.

messenger ribonucleic acid: molecule of ribonucleic acid that encodes a linear array of instructions for assembly of a protein. Messenger ribonucleic acid is transcribed from the DNA template, which is synonymous with a gene. This process is sometimes referred to as the central dogma of molecular biology.

multiple system atrophy (MSA): neurodegeneration caused by α-synuclein prions. Autonomic dysfunction, parkinsonism, and ataxia are prominent features. Cytoplasmic inclusion bodies composed of α-synuclein are widespread but are found primarily in the cytoplasm of oligodendrocytes, specialized brain cells that produce myelin. Olivopontocerebellar atrophy (OPCA), Shy-Drager syndrome (SDS), and striatonigral degeneration (SND) are all thought to be MSA variants.

mutation: alteration in DNA.

myoclonus: brief, involuntary twitching of a muscle or a group of muscles. Myoclonic jerks may occur alone or in sequence. Normal myoclonic jerks occur while falling asleep and do not signal a pathological process. In contrast, myoclonus at other times occurs in a wide variety of nervous system disorders.

nanometer: unit of length equal to one billionth of a meter (10^{-9} m).

neurodegenerative disease (ND): progressive deterioration of the nervous system due to abnormal protein processing or metabolism. Neurodegeneration causes neuronal dysfunction and eventually death.

neurofibrillary tangles: cytoplasmic inclusions within neurons composed of filaments of tau prions.

nuclear magnetic resonance spectroscopy: technique to determine the physical and chemical properties of atoms that are contained within particular molecules. Nuclear magnetic resonance can provide detailed information about the structure, dynamics, reaction state, and chemical environment of molecules.

nuclease: enzyme that catalyzes the hydrolysis of nucleic acid into nucleotides or oligonucleotides by cleaving phosphodiester linkages.

nucleic acid: general term for ribonucleic acid (RNA) and deoxyribonucleic acid (DNA), both of which are linear (unbranched) chains of nucleotides in which the 5′-phosphoric group of each one is esterified with the 3′-hydroxyl of the adjoining nucleotide.

oligodendrocyte: specialized brain cell that produces a myelin sheath that surrounds axons. These sheaths act as insulation to prevent the short-circuiting of electrical impulses and to speed up the transmission of signals along the axons of nerve cells.

paired helical filaments (PHF): paired filaments composed of tau prions. The individual filaments are 10 nm in diameter and make one full turn every 160 nm. They exhibit the tinctorial properties of amyloid and coalesce to form neurofibrillary tangles.

Parkinson's disease (PD): neurodegenerative disease caused by α-synuclein prions that assemble into fibrils that coalesce into threads and Lewy bodies. Synuclein prions cause dysfunction of dopamine neurons in the substantia nigra of the midbrain, producing pill-rolling tremors at rest, cogwheeling rigidity, slowed movements, difficulty walking, and masklike facies. Cognitive and behavioral problems may arise, with dementia commonly occurring in the advanced stages of the disease. PD is more common in older people, with most cases occurring after age fifty.

particle to infectivity (P/I) ratio: number of infectious pathogen particles in a single infectious unit. In the most efficient infections, the P/I ratio approaches unity.

pathogenesis: mechanism by which a disease is caused.

PCR (polymerase chain reaction): enzymatic method for the repeated copying of the two strands of DNA of a particular gene sequence. It is widely used to amplify minute quantities of DNA to produce quantities suitable for laboratory study, including sequencing.

pellet: fraction found at the bottom of a container after settling by gravity ($1 \times g$) or centrifugal force.

peptide: short segments of amino acids linked by peptide bonds. They are distinguished from proteins on the basis of size, typically containing fewer than fifty amino acids. The shortest peptide is a dipeptide consisting of two amino acids, also called residues, joined by a single peptide bond.

Pick body: inclusions within the cytoplasm of neurons. These inclusions contain aggregates of tau prions.

Pick's disease: rare neurodegenerative disease that causes progressive destruction of nerve cells in the brain. Symptoms include aphasia manifested as diminished speech as well as progressive dementia. Pick bodies filled with tau prions are found within neurons.

polymer: large molecule (macromolecule) composed of repeating structural units. These subunits are typically connected by covalent chemical bonds.

polymorphism: occurrence of something in several different forms, in particular two or more common amino acids at a single position in a sequence. The expression of methionine and valine at position 129 in human PrP represents a polymorphism.

posttraumatic stress disorder (PTSD): severe anxiety disorder that can develop after psychological trauma or a physically inflicted traumatic brain injury (TBI). Symptoms of PTSD include flashbacks or nightmares, avoidance of stimuli associated with the trauma, insomnia, anger, and hypervigilance. Such symptoms persist for more than one month and cause significant impairment in social and occupational activities. Some cases of PTSD worsen, with depression, apathy, and difficulties finding words; alcohol and drug abuse are common in patients with PTSD and suicide may also occur.

prion: infectious protein. Normal proteins refold into alternative structures called prions—this process becomes self-propagating. Most prions are rich in beta-sheet structure. Some prions participate in the normal cellular processes, while others feature in the pathogenesis of disease.

prion disease: disorder caused by prions. Such diseases can manifest as genetic, infectious, or sporadic maladies.

prion protein (PrP): protein of ~250 amino acids encoded by a chromosomal gene in mammals and birds. The normal, cellular form (PrP^C) is converted into the disease-causing form (PrP^{Sc}) by an as yet undefined process whereby PrP undergoes a profound conformational change.

Prnp: gene that codes for the prion protein (PrP), located on syntenic chromosome 2 in mice. _Prnp_ controls the length of the prion incubation time and is congruent with the incubation-time genes _Sinc_ and _Prni_.

PRNP: gene that codes for the human prion protein, located near the end of the short arm of chromosome 20.

progressive supranuclear palsy (PSP): neurodegeneration caused by tau prions. A majority of cases present with balance difficulties, personality changes, slowed movements, and impaired vertical gaze. Later, dementia becomes prominent, with disinhibition and diminished executive function as well as slurred speech and difficulty swallowing. Neuropathologically, inclusions of straight filaments of tau within astrocytes are seen. Originally called Steele-Richardson-Olszewski syndrome.

protease: enzyme that is used to catalyze the hydrolysis of peptide bonds of proteins.

protein: action molecules of life; they consist of long chains of amino acids that are covalently linked. Enzymes, antibodies, cytokines, and some hormones are examples of proteins. The amino acid sequences of proteins are encoded in genes. Three DNA base pairs encode one amino acid.

proteinase K (PK): enzyme (protease) that is used to catalyze the hydrolysis of a protein. Proteinase K is obtained from the fungus *Tritirachium album,* which derives its carbon and nitrogen from keratin, to which the letter K refers.

PrP amyloid rods: rod-shaped particles composed of PrP 27–30 that are indistinguishable from other amyloids both ultrastructurally and tinctorially. PrP amyloid rods possess high levels of prion infectivity. Also called prion rods.

PrP^C: normal, cellular isoform of the prion protein. PrP^C has three α-helices and two very short β-strands. PrP^C is readily digested by proteases.

PrP^{Sc}: alternatively folded isoform of the prion protein that is the sole component of the infectious PrP prion particle. This protein is the only identifiable macromolecule in purified preparations of prions. PrP^{Sc} has a high β-sheet content. Limited digestion of PrP^{Sc} leads to amino-terminal truncation, producing PrP 27–30. The "Sc" superscript was originally derived from "scrapie" but is now used generically to indicate all disease-causing isoforms of PrP.

PrP 27–30: fragment of PrP^{Sc} of ~140 amino acids, generated by amino-terminal truncation through digestion with proteinase K. The numbers refer to the molecular mass range (in kilodaltons) of this PrP fragment. PrP 27–30 retains prion infectivity and polymerizes into amyloid.

reagent: substance added to a solution of another substance to participate in a chemical reaction.

RNA (ribonucleic acid): chain of nucleotides in which the 5′-phosphoric group of each one is esterified with the 3′-hydroxyl of the adjoining nucleotide. RNA is found in both nuclei and cytoplasm of all cells. Messenger RNA contains the instructions for the assembly of amino acids into proteins.

scrapie: communicable prion disease in sheep and goats manifest as pruritus, abnormalities of gait, and invariably death.

scrapie-associated fibrils (SAFs): fibrils composed of unknown composition that are found in tissue homogenates prepared from animals with scrapie and patients who died of CJD. SAFs are composed of two or four filaments that are 12–16 nm in diameter and narrow to 4–6 nm every 40–60 nm, though some exhibit a periodicity of 80–110 nm. SAFs are ultrastructurally and tinctorially distinct from amyloid.

SDS: sodium dodecyl sulfate, a detergent used to denature proteins by disrupting hydrophobic interactions. SDS is employed in the electrophoretic analysis of macromolecules such as proteins.

Shy-Drager syndrome: a form of multiple system atrophy.

slow virus: virus causing a disease, which is preceded by a long incubation period of months to years. HIV is a slow virus that causes AIDS; many years may pass between the initial HIV infection and the manifestation of immunodeficiency.

somatic mutation: alteration in DNA that occurs after conception. Somatic mutations can occur in any of the cells of the body except the germ cells (sperm and egg) and therefore are not passed on to children.

spinobulbar muscular atrophy (SBMA) or Kennedy's disease: X-linked, adult-onset motor neuron disease characterized by slowly progressive weakness of the bulbar and extremity muscles. SBMA is the first disorder shown to be caused by a polyglutamine expansion. This mutation in the androgen receptor protein causes gynecomastia in males and muscle weakness due to motor neuron malfunction.

spinocerebellar ataxia (SCA): autosomal dominant disorders characterized by slowly progressive gait difficulties as well as truncal and limb ataxia manifest by progressive loss of coordination of the extremities. Speech difficulties and abnormal eye movements are frequently found. Atrophy of the cerebellum generally occurs. Six different SCAs are caused by polyglutamine expansions in proteins called ataxins.

sporadic Creutzfeldt-Jakob disease (sCJD): most common form of CJD. sCJD appears to occur worldwide at a rate of approximately one case per million population. Most cases involve older adults.

sporadic disease: occurs randomly with no known genetic, infectious, or any other identifiable cause. Most common forms of neurodegenerative diseases are sporadic disorders that seem to occur spontaneously via a stochastic or random process. One possible explanation involves the generation of prions within the protein clearance pathways where a prion precursor protein adopts a β-sheet-rich shape, becomes resistant to breakdown, and undergoes self-propagation. Another scenario involves somatic mutations that lower the threshold for a particular protein transforming into a prion. Once somatic mutations generate a sufficient number of prions to set up a self-propagating process, prion infection ensues.

strain: distinct variety of a pathogen that results in a characteristic phenotype. Strains of viruses, bacteria, and fungi can differ by as little as one DNA base pair that results in a single amino acid substitution in a protein encoded by one of these pathogens. Polio vaccines confer immunity to three different strains of polioviruses.

strain of prion: distinct variety of prion that produces a characteristic phenotype, as measured by clinical characteristics and/or the distribution of lesions. Much evidence argues that the biological information harbored by a particular prion strain is enciphered in the conformation of the protein. For PrP^{Sc} prions, strains can modify the length of the incubation period, the distribution of PrP^{Sc} in the brain, and the pattern of neuropathological lesions. Different strains of tau prions appear to be responsible for the different primary tauopathies that include progressive supranuclear palsy, corticobasal degeneration, Pick's disease, behavioral variant of frontotemporal dementia, and argyrophilic grain disease.

sucrose gradient: technique used to separate macromolecules or larger particles based on their size and/or density. Decreasing concentrations of sucrose are layered in a centrifuge tube and the sample generally added at the top. Upon centrifugation, particles travel through the gradient as a function of size and density.

supernatant: fraction found at the top of a container after settling by gravity $(1 \times g)$ or centrifugal force.

tauopathy: neurodegenerative disease caused by tau prions. Primary tauopathies include inherited and sporadic forms of frontotemporal dementia (FTD), including progressive supranuclear palsy, corticobasal degeneration, Pick's disease, behavioral variant of FTD, and argyrophilic grain disease. Secondary tauopathies include Alzheimer's and postencephalitic Parkinson's diseases, as well as rabies, Niemann-Pick C disease, and some familial forms of Creutzfeldt-Jakob and Gerstmann-Sträussler-Scheinker diseases. Dementia pugilistica and chronic traumatic encephalopathy are also secondary tauopathies.

tau protein: microtubule-associated protein that can become a prion and accumulate in neurons as neurofibrillary tangles or Pick bodies.

transgenic (Tg): referring to an organism in which nonnative DNA has been introduced into the germ cells by injection into the nucleus of the ovum.

traumatic brain injury (TBI): intracranial injury that occurs when an external force damages the brain. TBIs are commonly subdivided into mild, moderate, and severe, with the duration of lost consciousness a key factor in ranking severity.

titer: concentration of infectious particles as determined by the maximum dilution at which a preparation is still infectious.

titration: determination of the concentration of an infectious pathogen by serial dilution until the last dilution does not give a positive test. This value is called the titer.

tobacco mosaic virus (TMV): cylindrical virus that infects plants, especially tobacco and other members of the Solanaceae plant family. A protein shell surrounds a core of viral RNA.

transgene: gene from one organism that has been transferred and integrated into the DNA of another organism of the same or different species, such that the transferred gene is expressed in the new host organism. Investigators conducting prion transmission studies use transgenes to convey unnatural molecular characteristics to experimental animals so as to circumvent the species barrier. For example, transmission of hamster prions to transgenic mice expressing hamster PrP abrogates the species barrier. When the mouse *Prnp* is disrupted (*Prnp$^{0/0}$*) in transgenic mice expressing hamster PrP, the incubation period for hamster prions decreases. In transgenic mice expressing human PrP, disruption of the mouse PrP gene is required for transmission of human prions.

transmissible mink encephalopathy (TME): prion disease in mink.

transmissible spongiform encephalopathy (TSE): general term for neurodegeneration caused by PrP^{Sc} prions. The term "TSE," which predates the term "prion disease," is ambiguous with respect to both etiology and neuropathology: TSE emphasizes the infectious form of prion disease and ignores the much more common genetic and sporadic forms. Moreover, the spongiform changes in the CNS of mammals harboring PrP^{Sc} prions can be quite variable, with little vacuolation sometimes seen.

vacuole: minute space in the brain that is filled with aqueous fluid and often accompanies neurodegeneration. Some strains of PrP^{Sc} prions stimulate vacuolation while others do not. Intense vacuolation ranges from spongiform degeneration to status spongiosis.

variant Creutzfeldt-Jakob disease (vCJD): prion disease in young adults who were exposed to BSE prions in Europe or other high-risk countries. Florid plaques are pathognomonic of vCJD; such PrP amyloid plaques are surrounded by large vacuoles. Most vCJD cases have occurred in young adults. The only risk factor for vCJD was the consumption of hamburgers in Great Britain more than two times per week.

virion: complete virus particle that is structurally intact and infectious.

viroid: infectious pathogen of plants that is smaller than a virus and consists only of single-stranded closed circular RNA, lacking a protein covering (capsid); replication does not depend on a helper virus but is mediated by host cell enzymes. The molecule weight of viroids is 75–100 kDa.

virus: infectious pathogen that is capable of passing through fine filters that retain most bacteria, are not visible in the light microscope, lack independent metabolism, and are incapable of growth or reproduction apart from living cells. Virus particles contain either DNA or RNA genomes that direct the synthesis of progeny viruses. A protein shell or capsid protects the viral nucleic acid. Viruses range in size from 15 nm to several hundred.

wild-type: organism without mutations; the phenotype that is naturally occurring or characteristic of most members of the species.

Acknowledgments

The preparation of this book was a long labor, and I wish to acknowledge the special contributions of Fred Cohen, Stephen DeArmond, Martin Gilbert, Stephen Hauser, Michael Kahn, Susan Kahn, Colleen Leof, David Leof, Hang Nguyen, Oliver Sacks, Jean Thomson Black, and Herman Wouk. For their thoughtful critiques of the manuscript, I am extremely grateful. Pinning down my family origins was a monumental task, which I could not have done without the help of Shari Birnbaum Baum, Ethel Baron Berlau, Harold Birnbaum, Susan Bulbin, David Conway, Yuri Dorn, Charles Feldman, Alan Glynn, Ian Glynn, Nancy Jacobson, Kate Levine, William Mednick, Chris Prusiner, Paul Prusiner, and James Spigel. Their contributions were invaluable in helping me to verify people and events as well as to identify inaccuracies in this unrecorded history.

I am very appreciative of the opportunities given to me by the American people, who through their generosity, wisdom, and taxes have supported our scientific investigations for more than three decades. Support from the NIH, DOD, and NSF, as well as from the State of California, has been crucial. Equally important has been support from the Stephen D. Bechtel Jr. Foundation, Clausen Family Foundation, Community Foundation Silicon Valley, Cure PSP Foundation, Dana Foundation, Eagleson Institute, Sherman Fairchild Foundation, John Douglas French Alzheimer's Foundation, Paul Galvin Memorial Foundation, Glenn Foundation for Medical Research, Larry L. Hillblom Foundation, Howard Hughes Medical Institute, W. M. Keck Foundation, Anna-Maria and Stephen Kellen Foundation, Koret Foundation, Lincy Foundation, G. Harold and Leila Y. Mathers Foundation, Bernard Osher Foundation, Rainwater Charitable Foundation, R. J. Reynolds Company, Rosenberg Family Foundation, Sanofi-Aventis LLC, Stephen C. and Patricia A. Schott Foundation, and Taube Family Foundation.

To the extraordinary people who are the stewards of these wonderful foundations, including Wally Burke, Walter Burke, and Bonnie Himmelman; Mark Collins; Everett Cook; Laurie Dachs; Dennis Eagleson; Jim Handelman; and Jeff Wilkins, I am very appreciative.

I want to give special thanks to the generous people who have made gifts to support our work: Richard Blum and Dianne Feinstein; William and Ute Bowes; Cheryl Breetwor-Evans; Delia Brinton, Mary Jane Brinton, William and

Gerry Brinton, and Gregg and Katherine Crawford; Carmen Castro; Ron and Gayle Conway; Robert and Mary Galvin; Agnes and Berel Ginges; Andrew Gundlach; Clarence Heller; Christina and Michael Homer; Michael Kellen; Kerk Kerkorian; Bill McCabe; Gladyne Mitchell; John Osterweis; Richard Rainwater; Detlev and Hannelore Riesner; Marian and Herbert Sandler; and Norman Schultz.

Many scientific colleagues were so important at many points in the investigations. I am grateful to my mentors and teachers, who include J. Richard Baringer, Britton Chance, Robert Fishman, C. P. Lee, Olle Lindberg, Donald Macrae, R. Curtis Morris, Rudi Schmid, Lloyd (Holly) Smith, Louis Sokoloff, Earl Stadtman, and Sidney Wolfson.

Thanks also to my former and current students: Misol Ahn, Ronald Barry, Ilia Baskakov, Carolyn Bellinger-Kawahara, Paul Bendheim, David Berry, Thomas Blochberger, Jeffery Bockman, David Bolton, David Borchelt, Patrick Bosque, Kevin Boylan, Dale Bredesen, Darel Butler, Ed Choi, David Colby, Carola Doak, Doug Ehresman, Ruth Gabizon, Jean-Marc Gabriel, David Garfin, Maria Gasset, Michal Geva, Sina Ghaemmaghami, Rolf Hecker, Nick Hogan, Karen Hsiao, Zwei Huang, Kiyotoshi Kaneko, Zoltan Kanyo, Marcela Karpuj, Ken Kasper, Klaus Kellings, Aleks Kijac, David King, Carsten Korth, Hans Kretzschmar, In-Su Lee, Dan Lowenstein, Duo Lu, Richard Mamelok, James Mastrianni, Yoichi Matsunaga, Barney May, Michael McKinley, Ingrid Mehlhorn, Zeev Meiner, Zach Mensinger, Rudi Meyer, Richard Moore, Tamaki Muramoto, Karah Nazor, Bruno Oesch, Julian Ollesch, Keh-Ming Pan, David Peretz, Veronique Perrier, Puay Phuan, Lenore Pierera, John Price, Alex Raeber, Mark Rogers, Chongsuk Ryou, Herman Schätzl, Martin Schlumpberger, Michael Scott, Serena Spudich, Neil Stahl, Johannes Stöckel, Jan Stöhr, Surachai Supattapone, Erdem Tamgüney, Albert Taraboulos, Jörg Tatzelt, Glenn Telling, Eric Turk, Martin Vey, Rachel Wain, Joel Watts, David Westaway, Holger Wille, Fruma Yehiely, Hong Zhang, and Laurence Zulianello.

To my many collaborators whose numerous contributions made the scientific discoveries described here possible: Michael Alpers, Harry Baker, Michael Baldwin, Hayden Ball, J. Richard Baringer, Henry Baron, Leslie Benet, Herman Bujard, Al Burlingame, Dennis Burton, George Carlson, Mark Chatigny, James Cleaver, Tim Crow, Gustav Dallner, William DeGrado, Theodor Diener, Ken Downing, Raymond Dwek, Jane Dyson, Carl Eklund, Tamao Endo, Robert Fletterick, Pierluigi Gambetti, Michael Geschwind, Jason Gestwicki, Bernardino Ghetti, Kurt Giles, David Glidden, Joe Guglielmo, William Hadlow, John Hearst, Thomas Hökfelt, Leroy Hood, Matt Jacobson, Tom James, Ellis Kempner, Steve Kent, David Kingsbury, Akira Kobata, Krister Kristensson, Giuseppe Legname, Pierre Lessard, Emil Lin, Vishwanath Lingappa, Jeff Long, Elio Lugaresi, Susan

Marqusee, Lefkos Middleton, Bruce Miller, Michael Miller, Glenn Millhauser, William Mobley, Warren Olanow, Jurg Ott, Frank Owen, Daniel Perl, Peter Peters, Adam Renslo, Jesús Requena, Rosalind Ridley, Detlev Riesner, Gareth Roberts, Jiri Safar, Andrej Šali, Dan Serban, Brian Shoichet, Michael Silber, Daniel Stites, Gerald Stubbs, Jun Tateishi, Patrick Tremblay, Jonathan Weissman, Charles Weissmann, Jim Wells, Robley Williams, Anthony Williamson, and Peter Wright.

To my University of California colleagues, who provided so much support and guidance over decades: Arthur Asbury, John DeLuca, Susan Desmond-Hellmann, Ivan Diamond, Howard Fields, Juan Korenbrot, Julius Krevans, Joseph Martin, David Ramsay, Max Redfearn, Lou Reichardt, Howard Shapiro, Frank Sooy, Bruce Spaulding, Robert Stroud, Phil Weinstein, and Mark Yudof.

To my scientific colleagues at other institutions whose help at many different points in this scientific odyssey was invaluable: Oded Abramsky, Michael Brown, Jack Dixon, Lynn Enquist, Lars Ernster, Robert Gallo, Donald Gilden, Joe Goldstein, François Jacob, Richard Johnson, Ronald Kaback, Hilary Koprowski, Daniel Koshland, Edith Langner, David Lowenberg, Erik Lycke, Colin Masters, Maclyn McCarty, Neal Nathanson, Erling Norrby, Michael Oldstone, Fred Seitz, Michael Shelanski, Frank Westheimer, Charles Yanofsky, and Vincent Zigas.

To some very special members of my scientific and administrative staff: Greta Boesel, Howard Booth, Karen Bowman, Michael Braunfeld, Patricia Cochran, Robert Cotter, Fran Elvin, Yevgeniy Freyman, Lorraine Gallagher, Steve Glaser, Darlene Groth, Cornelia Heinrich, Brenda Klein, Ruth Kohler, Oanh Nguyen, Panos Philandrinos, Ana Serban, Allison Taylor, and Monika Wälchli.

Last, my gratitude to some special administrators at the NIH for their enthusiastic support over many years: Robert Butler, Murray Goldstein, Richard Hodes, Al Kerza-Kwiatecki, Zaven Khachaturian, Carl Levinthal, Andrew Monjan, Steve Snyder, Remesh Vermuri, and Franklin Williams.

Illustration Credits

Figure 1: Courtesy of Bill Mednick and Rick Gilinsky, and from personal family archives

Figure 2: From personal family archives

Figure 3: Courtesy of Sidney Wolfson, Olle Lindberg, Earl Stadtman, Britton Chance, and Charles Yanofsky

Figure 4: Photomicrograph prepared by Stephen DeArmond

Figure 5: Courtesy of Ivan Diamond

Figure 6: Reprinted, with permission, from S. Hornsey and J. Denekamp, "Tikvah Alper: An indomitable spirit, 22 January 1909–2 February 1995," *Int. J. Radiat. Biol.* 71: 631–42 (1997)

Figure 7: Courtesy of Rocky Mountain Laboratories, NIAID

Figure 8: Photo of Dickinson and Kimberlin reprinted with permission from Niall Cotton; photo of Marsh courtesy of the University of Wisconsin, Madison, archives

Figure 9: Reprinted, with permission, from D. C. Gajdusek, "Observations on the early history of kuru investigations," in *Slow Transmissible Diseases of the Nervous System,* vol. 1, ed. S. B. Prusiner and W. J. Hadlow (New York: Academic Press, 1979), 7–36

Figures 10, 11, 27: From the author's personal collection

Figure 12: Courtesy of Jack Seitz, Ingbert Gruttner/Rockefeller University, Curtis Morris, and The Regents of the University of California

Figure 13: Reprinted, with permission, from S. B. Prusiner, "Prions," *Sci. Am.* 251: 50–59 (1984)

Figure 14: Courtesy of the Department of Chemistry at Harvard University

Figure 15: From W. L. Buller, *A History of the Birds of New Zealand* (London, 1888)

Figure 16: Courtesy of G. Paul Bishop, Jr., used with permission

Figure 17: Courtesy of Ted Diener

Figure 18: Courtesy of Barbara Horgan of The George Balanchine Trust

Figure 19: From the author's personal collection, provided by Bonnie Himmelman

Figure 20: Micrograph of prion rods reprinted with permission from M. P. McKinley, M. B. Braunfeld, C. G. Bellinger, and S. B. Prusiner, "Molecular characteristics of prion rods purified from scrapie-infected hamster brains," *J. Infect. Dis.* 154: 110–20 (1986); micrograph of TMV prepared by Amy Kendall and Gerald Stubbs

Figure 21: Immunoblot adapted from S. B. Prusiner, "Prions (Les Prix Nobel Lecture)," in *Les Prix Nobel,* ed. T. Frängsmyr (Almqvist and Wiksell International, Stockholm, 1997), 268–323. Photos courtesy of Ron Barry and Stephen DeArmond

Figure 22: Courtesy of the Institute for Systems Biology

Figure 23: Courtesy of Charles Weissmann

Figure 24: Courtesy of Detlev Riesner

Figure 25: Courtesy of the McLaughlin Research Institute and Neil Stahl

Figure 26: Courtesy of Bruce Miller/UCSF Memory and Aging Center and Karen Hsiao Ashe

Figure 28: Courtesy of Neil Stahl and Fred Cohen

Figure 29: Courtesy of Lord John MacGregor

Figure 30: Courtesy of Glenn Telling

Figure 31: Reprinted with permission from R. B. Setlow, "Shedding light on proteins, nucleic acids, cells, humans and fish," *Mutat. Res.* 511: 1–14 (2002)

Figure 32: Courtesy of The Nobel Foundation

Figures 33, 37: Micrographs courtesy of Stephen DeArmond

Figure 34: Drawing courtesy of Stephen DeArmond

Figure 35: Courtesy of Mathias Jucker and Lary Walker

Figure 36: Courtesy of Zachary Mensinger

Index

Page numbers in **boldface** refer to illustrations.

sequence paper, 126, 131–33, 136–38, 144, 151; Weissmann's paper on knockout mice, 180–82

Centers for Disease Control (CDC), 255

Central Veterinary Laboratory, Weybridge, Surrey, 196

centrifugation, 37–43, 65, 67, 86, 143

cerebellum, 52, 62–63; damage to, 62–63, 106; spongiform, 106

Chance, Britton, 12, **13**

Chandler, Richard, 26

Chatigny, Mark, 41–42, 69

chemistry, 1–2, 31

Chesebro, Bruce, 137, 142, 162, 172, 177

Chicago, 3

chimpanzees, 34, 103; CJD transmitted to, 50, 56–57, 154; kuru transmitted to, 52–54, 93

Christen, Yves, 232

Christopher Columbus Quincentennial Discovery Award, 179

chronic traumatic encephalopathy (CTE), xvi. *See also* frontotemporal dementias (FTDs)

Churchill, Winston, xvii

Ciba Foundation, 150

Cincinnati, 1–2, 8, 10, 11

Cleaver, Jim, 210

Clements, John, 185

cloning, 65, 129–39, 144, 156–57

C9orf72 gene, 246, 247

Cochran, Patricia, 67, 68

coding mutation, 161

codons, 157, 159, 161, 169, 209, 270n6

Cohen, Fred, 171–72, **172**, 173–74, 217, 243

Colby, David, 243

Cold Spring Harbor Laboratory, Long Island, 33, 124

colinearity, 22, 25

Collinge, John, 169

Columbia University, 106, 118–19, 157, 249

competition, scientific, 46–49, 115, 119–24, 133–39, 144–52, 158–68, 179–83, 184–92, 215–17

Compton Research Institute, 30, 33, 37, 43, 49, 163

conformation, 25, 217

Congo red dye, 106, 113, 119, 120, 122, 133

Congress, 69

coprion, 168

corticobasal degeneration (CBD), 246

Creutzfeldt, Hans, 56–57

Creutzfeldt-Jakob disease (CJD), xvi, 17–24, 30, 35, 56–57, 63, 103–4, 149, 153–69, 193, 221, 232, 233, 238, 245, 249, 250; amyloids and, 108–24; ataxic, 106–7, 154; brain, 17–18, **18**, 19–24; -BSE link, 193–200; familial, 153–55, 160, 168, 169, 207–9, 233, 234; genetics of, 153–69; in George Balanchine, 105–7, 113, **117**, 118; growth hormone–induced, 200–204; kuru and, 56–57; prion hypothesis, 93; -scrapie connection, 41, 56–57, 104; sporadic, 160–61, 169, 206–7, 234; terminology, 56–57; transmission to chimpanzees, 50, 56–57, 154; variant, 193–94, 223; viral hypothesis, 17–24, 29–35, 94, 122, 149, 158–68, 189, 203, 206, 221

Crow, Tim, 156, 157, 169

C-terminus, 25, 126

CT scan, 102, 105

cysteines, 211, 212

cytosine, 24

MacGregor, John, 196–97, **197**, 198–200

Macleod, Colin, 75, 96

MacNeil/Lehrer NewsHour, 115

Macrae, Donald, 18, **19**, 74, 76

mad cow disease. *See* bovine spongiform encephalopathy (BSE)

Maeterlinck, Maurice, 96

Manuelidis, Laura, 223–24

Marsh, Richard, **44**, 47–49, 65, 67, 81, 205–7; viroids and, 47–49

Martin, Joseph B., 214

Martin, William, 273n5

mass spectrometry (MS), 170–71

Masters, Colin, 154–55

Mastrianni, James, 209

McArthur, John, 51

McCarty, Maclyn, 75, **76**, 96, 185, 186

McKinley, Michael, 100–101, 103, 110, 113, 123

measles, 33

Meiner, Zeev, 216

Meitner, Lise, 19

membrane fragments, 43–46

memory prion, 249–50

meningoencephalomyelitis, 21

Merz, Patricia, 119, 151

metabolism, 12, 175; brown fat, xv, 2, 12, 13–14, 22; glutamate, 23

mice, 26, 37, 104, 140–42, 169; genetic control of incubation times in, **139**, 140–42; GSS, 158, 167; ILnJ, 140–42, 268n10; knockout mouse experiments, 176–83, 215–16; NZW, 140–42, 268n10; scrapie experiments, 26–49, 65, 68–69, 116–18, 140–42, 151–52, 169, 176–83; scrapie transmission from mouse to mouse, 26–35; transgenic,

162–69, 176, 179, 207–9, 235–42, 248–49, 250; transmission of prions between hamsters and, 162–69; VM/Dk, 113, 116–18, 140, 142

Middleton, Lefkos, 249

Miescher, Friedrich, 96

military, brain injuries in, 244, 245, 258–59

Miller, Bruce, 155–56, **156**, 157

MIT, 187

monkeys, 54, 154, 158, 201, 238, 249

Morris, Curtis, 75, **76**, 79, 85

Moscow, 3, 6

MRI, 102

mRNA (messenger ribonucleic acid), 24, 129, 134, 135–36

Mugnier, Elisabeth, 204

multiple sclerosis, 39, 45, 155, 189

multiple system atrophy (MSA), xiv, 248–49

mutation, 156–59, 208; GSS, 156–60, 167, 168; neurodegeneration and, 232–52; PrP, 156–61, 167–69, 208; types of, 161

nanometer, 28, 29, 45, 111

National Academy of Sciences, 179, 214

National Creutzfeldt-Jakob Disease Surveillance Unit (NCJDSU), 193–94

National Institutes of Health (NIH), xv, 14–15, 17, 21, 50, 53, 58, 114, 171, 190, 260; funding, 27, 32–35, 38–39, 59, 65–67, 71, 79, 108, 144, 177, 178; Institute of Allergy and Infectious Diseases, 65; Prusiner and, 14–15, 16, 32–35, 38–39, 65–67, 71, 79, 108, 144, 171, 177, 179; Virology Study Section, 32–33, 38–39

Pattison, Iain, 30, 162–63, 205
PCR (polymerase chain reaction), 157, 168–69
pellet, 37–43
peptides, 82, 116, 125
Perlman, David, 89–90, 91, 92, 143, 269*n*1
Pettersson, Ralf, 219, 226–27
phenol, **83**
phenotypes, 141, 209
Philadelphia, xv, 11, 86
phosphatases, 176
Pick's disease, 114, 246
pidgin, 59, 61
Pneumococci bacteria, 96
polio, 21, 33, 46, 99, 257
polymorphism, 169
posttraumatic FTD, 244. *See also* chronic traumatic encephalopathy (CTE)
posttraumatic stress disorder (PTSD), 244–45, 258–59
Potamkin Prize for Alzheimer's Disease and Related Disorders Research, 177
precursor proteins, 233–34, **234**, 236–44
pregnancy, 154
prion (bird), 88, **89**
prion protein (PrP), 81–82, **83**, 84–96, 97–107, 125–42, 166; amino acid sequence of, 126–39, 144, 171–75; conversion of PrP^C into PrP^{Sc}, 170–76; Gajdusek's discovery claims, 189–91; gene probe, 125–26; knockout mouse experiments, 176–83, 215–16; mutation, 156–61, 167–69, 208; neurodegeneration and, 232–52; strain issue, 205–11; structure, 170–83; transgenes, 162–69, 176, 207–9, 235–42, 271*n*10

prion rods, 110–11, **112**, 113–24, 127–29, 132–33
prions, xvi-xviii, 15, 85–93; amyloid connection, 108–24; BSE, 193–200; *Cell* paper on, 102–3; collapse of opposition to, 161–62, 181; competition, 46–49, 50, 115, 119–24, 133–39, 144–52, 158–68, 179–83, 184–92, 215–17; conversion of PrP^C into PrP^{Sc}, 170–76; *Discover* attack on, 144–52, 225; discovery of, xvi, 85–95; genetic origin of PrP *27–30*, 125–42; genetics of, 125–42, 153–69; growth hormone–induced CJD, 200–204; human disease, 153–69; hybrid, 166; hypothesis, 85–107; knockout mouse experiments, 176–83, 215–16; leveraging the prion revolution, 259–60; neurodegeneration and, 232–52; Nobel Prize for discovery of, 213–31; nucleic acids and, 81–96, 118, 130–31, 137–38, 145–46; opposition to hypothesis, 93–96, 119–24, 134, 143–52, 158–68, 185–91, 222–25; *Science* article on, 85–93, 97, 100, 145; strains, 152, 205–11; synthetic, 212, 244–46; terminology, 86–95, 97, 103, 144–48; transgenes, 162–69, 176, 207–9, 235–42, 271*n*10; variations in shape, 205–11; viral hypothesis, 17–24, 29–35, 39–40, 43–49, 94, 122, 134–38, 144–49, 160, 162–69, 187–91, 203, 206, 221, 267*n*15. *See also specific prions and prion diseases*
Proceedings of the National Academy of Sciences, 40, 95, 110; "Viroids and Prions," 95

progranulin, 246, 247

progressive supranuclear palsy (PSP), 246

proline, 25, 157, 161

proteases, 47, 82, **83**, 84, 100, 101, 128, 136, 170, 188, 207, 208

proteinase K (PK), 189

proteins, 24–25, 26, 28–29; amino acid sequence, 126–39, 144, 171–76; amyloid, 108–24; definition of, 24–25; electrophoresis, 97–101; folded, 171, 175–76, 206, 217; identification, 97–107; infectious, 81–107, 201, 213; neurodegeneration and, 232–52; plasticity and alternative folding, 175–76; prion, 81–82, **83**, 84–96, 97–107, 136, 166; PrP 27–30, 103–5, 125–42; size of, 99; structure of, 170–83. *See also* prion protein (PrP); *specific proteins*

provirus, 29, 269*n*6

PrP. *See* prion protein (PrP)

PrP^C, 117, **117**, 136, 144, 160, 170–83, 199, 217; discovery of, 136; knockout mouse experiments, 176–83; purification of, 173–74; strain issue, 205–11; structure of, 170–83; transgenes, 162–69, 176, 207–9

PrP^{Sc}, 117, **117**, 136, 160, 170–83, 199, 245, 250; discovery of, 136; hypotheses for spontaneous formation of, 160–61; strain issue, 205–11; structure of, 170–83; transgenes, 162–69, 176, 207–9

PrP *27–30,* 103–4, 125–42; amino acid sequence of, 126–39, 144, 171–76; -amyloid connection, 108–24; *Cell* paper on, 126, 131–33, 136–38, 144,

151; conversion of PrP^C into PrP^{Sc}, 170–76; Gajdusek's discovery claims, 189–91; genetic origin of, 125–42; SAFs and, 119–24, 132–33; strain issue, 205–11; structure of, 170–83; terminology, 103–4

Prusiner, Ben, **4**, 6, 9

Prusiner, Bessie, **4**

Prusiner, Dave, **4**

Prusiner, Ethel Galinsky, **4**, 6, 10, 11

Prusiner, Feine, 3, **4**, 5, **5**, 6

Prusiner, Lawrence Albert, 3, **4**, **5**, 6, 8–11, 24

Prusiner, Lippman-David, **4**, 6

Prusiner, Miriam Spigel, 3, **5**, 7, 8–11, 24

Prusiner, Molly, **4**

Prusiner, Noach, 6

Prusiner, Paul, **4**, 9

Prusiner, Sandy, 219

Prusiner, Stanley Arnold, **4**, 6, 8

Prusiner, Stanley B., **4**, **5**, **228**; amyloid/prion connection, 108–24; ancestry and upbringing of, 1–15; awards and honors, 177, 180, 184–88, 213–31; beginnings of CJD research, 16–25; board certification in neurology, 74; *Cell* paper on amyloids, 114–16; *Cell* paper on prions, 102–3; *Cell* paper on PrP 27–30 sequence, 126, 131–33, 136–38, 144, 151; collapse of opposition to, 161–62, 181; competition and, 46–49, 50, 115, 119–24, 133–39, 144–45, 158–68, 179–83, 184–92, 215–17; criticism of, 70–74, 90–96, 119–24, 134, 143–52, 158–68, 185–91, 214–15, 221–25; *Discover* attack on, 144–52, 225; education of, 1–2, 11, 12–14;

Prusiner, Stanley B. (continued)
falling out with Weissmann, 180–83; fame of, 184–88, 213–31; family of, 219; funding issues, 27, 32, 34–35, 59, 65–67, 71–80, 108–10, 144, 147, 165, 167, 177, 178; Hadlow collaboration, 33–35, 38, 65–67; health problems of, 102–3; as Hughes investigator, 35, 71–73, 75, 76; introduction of prion hypothesis, 85–95; lab issues, 32–35, 41, 66–71, 74–80, 110, 123–24, 167; Lasker Award, 184–85, 213; *Nature* paper on GSS, 158–59; New Guinea kuru investigation, 57–64; *New Yorker* article on, 223–24; NIH and, 14–16, 32–35, 38–39, 65–67, 71, 79, 108, 144, 171, 177, 179; Nobel Prize, 213–31; opposition to, 70–74, 90–96, 119–24, 134, 143–52, 158–68, 185–91, 222–25; prion hypothesis, 85–107; publicity and, 143–52, 184–88, 213–31; *Science* article on prions, 85–93, 97, 100, 145; scientific mentors of, 12–15; on "strain" issue, 205–11; tenure battle, 28, 65–80; Tg mouse studies, 162–69, 207–9, 235–42; at UCSF, 14, 18, 23, 26, 32, 35, 39, 65–80, 171, 213–14
Prusiner, Wulf, 3, **4**, 5, **5**, 6, 7
psoralen photoadducts, **83**
purification, 28, 31, 126–29; of PrP^C, 173–74; of scrapie agent, 28–31, 36–49, 81–85, 95, 100–102, 126–29, 189, 265n1

rabbits, 104, **117**, 178
radiation, 28–31, 43, 45, 93, 98, 210
radioactive iodine, 98–99, 103

Ramsay, David, 167
R.C. (patient with CJD), 154
Reagan, Nancy, 254–55
Reagan, Ronald, 254–55
Reynolds, R. J., 75–76, 108
Rhodes, Richard, 222–24; *Deadly Feasts*, 214–15; *New Yorker* article on Prusiner, 223–24
Ridley, Rosalind, 156, 238
Riegler, Rudolf, 215, 218
Riesner, Detlev, 138, **138**, 140, 217, 218
Ringertz, Nils, 218–19, 222
RNA (ribonucleic acid), xvii, 20, 24, 29–30, 37, 48, 81–84, 111, 130–31, 168, 205, 269n6; viroids, 94–95
RNases, 47, 82, 83, **83**, 171, 174–75, 217
Robert Koch Institute, Berlin, 120, 122
Roboz-Einstein, Elizabeth, 32
Rockefeller Institute, 96
Rockefeller University, 75
Rocky Mountain Laboratories, Hamilton, Montana, 33–34, 58, 65–66, 132, 137, 142
Rodbell, Martin, 213
Roentgen, Wilhelm, 229
Rohwer, Robert, 45, 187, 189
Roos, Raymond, 154
rotary shadowing, 111
Royal Society, London, 192
Russia, 3, 6; Jews, 3, 6, 19

Sacks, Helen Spigel, 11, 86
Sagan, Carl, 262n4
Salk, Jonas, 21, 33
Salk polio vaccine, 21, 33
San Francisco, xiv–xvi, 14, 49, 67, 69, 87, 123, 130, 136–37, 177, 178
San Francisco Chronicle, 89, 91, 92, 143, 269n1
Sarcosporidia, 29

Schaffer, Bernard Black, 2

Scheinker, Isaak, 155, 270*n*3

Science, 85, 90, 102, 205; "Novel Proteinaceous Particles Cause Scrapie," 85–93, 97, 100, 145; SAF paper, 121–23

Scotland, 21, 198

Scott, Mike, 162–63, **163**, 164–65, 167, 271*n*10

scrapie, 17, 18–49, 198, 232, 238, 243, 262*n*4, 268*n*4; amyloids and, 108–24; centrifugation experiments, 37–43, 65, 67, 86, 143; -CJD connection, 41, 56–57, 104; competition, 46–49, 50, 115, 119–24, 133–39, 144–52, 158–68, 179–83, 184–92, 215–17; hamster studies, 49, 65–69, 82, 94, 98–102, 110–13, 118, 130, 134–35, 162–68, 173; initial protocol for measuring scrapie agent, 27–28; knockout mouse experiments, 176–83, 215–16; -kuru connection, 33–34, 41, 53, 56, 93; membrane hypothesis, 43–46; molecular structure of, 43–46, 58; mouse studies, 26–49, 65, 68–69, 116–18, 140–42, 151–52, 169, 176–83; nucleic acids and, 81–96, 118, 130–31, 137–38, 145–46; prion hypothesis, 85–107; proteins, 81–82, **83**, 84–96, 97–107, 136, 166; PrP *27–30,* 103–5, 135–42; purification, 28–31, 36–49, 81–85, 94, 100–102, 126–29, 189, 265*n*1; race, 36–49; research funding, 27, 32, 34–35, 38–39, 59, 65–67, 71–80, 108–10, 144, 147, 165, 167, 177, 178; research labs, 32–35, 41, 66–71, 74–80, 110, 123–24, 167; theories and, re-

search, 26–35, 36–49, 65–96, 108–24; titration measurements, 26–32, 45, 65; transmission from mouse to mouse, 26–35; transmission from sheep to mice, 26; viral hypothesis, 29–35, 39–40, 43–49, 94, 122, 134–38, 144–49, 160, 162–69, 187–91, 203, 206, 221

scrapie-associated fibrils (SAFs), 119–24, 132–33; *Science* paper, 121–23

SDS (sodium dodecyl sulfate), **83**, 98, 208

Seitz, Fred, 75, **76**, 109, 147, 148; *The Modern Theory of Solids,* 109

Sell, Kenneth, 65–67

Semancik, Joe, 48–49

Serban, Ana, 210

Setlow, Richard, 210–12

Shannon, James, 75

Sharp, Phillip, 187

sheep, 104, 163, 197, 206, 237; BSE and, 197–98, 199; transmission of scrapie to mice from, 26

Shy, Milton, xiv, 248

Shy-Drager syndrome. *See* multiple system atrophy (MSA)

silent mutation, 161

Sioux City, Iowa, 3–6

Sixth International Congress of Virology (1984), 121

slow virus, 17, 19, 22, 32–33, 39, 43, 58, 86, 92, 145, 146, 224

Smadel, Joseph, 51, 53

Smith, Holly, 14, 71, 75, 77

snout and rooting reflexes, **60**, 63, 105

Society for Neuroscience, xvi

somatic mutation, 161

Sooy, Frank, 79, 80, 91, **92**

South Africa, 19

Southwood, Richard, 196
Southwood Working Party, 196
species barrier, 162–63
Spigel, Ben, **5**, 7, 8, 10
Spigel, Herbert, **5**, 11
Spigel, Miriam, **5**, 11
Spigel, Mollie Feldman, **5**, 7, 8–9, 11
Spigel, Moses Herman, 7
Spigel, Naomi, **5**, 11
Spigel, Sarah Weinstein, 7
spleen, 37–41
spongiform, 17, **18**, 21, 106, 154, 155,
    157, 168, 179; BSE, 17, 18, 151,
    193–200, 206, 207, 223, 232
sporadic Creutzfeldt-Jakob disease
    (sCJD), 160–61, 169, 206–7, 234
sporadic neurodegenerative diseases,
    233, 234, **234**, 235
Stadtman, Earl, **13**, 14–15, 23, 84, 171,
    186, 265*n*1
Stahl, Neil, 170–71, 174
Stanford University, 22, 235
Stites, Dan, 117
Stockholm, xiii-xv, 122, 214, 215–31
Stöhr, Jan, **239**, 240
strains, prion, 152, 205–11
strain typing, 152
Sträussler, Ernst, 155, 270*n*3
stroke, 257, 258
subacute sclerosing panencephalitis
    (SSPE), 33
subvirus, 30
sucrose gradient, 42, 110, 111
Südhoff, Tom, 274*n*1
sugar, 175; chains, 29
suicide, 259
supernatant, 37–43
superoxide dismutase *1*, 249
Sweden, 12, 13, 189, 215–31
Swedish Academy of Sciences, 215

synthetic prions, 212, 244–46
synuclein prions, 237–38, 247–50, **251**

Tamgüney, Erdem, 238
Taraboulos, Albert, 191
Tateishi, Jun, 158, 159
Taubes, Gary, 144, 224–25; *Discover*
    article on prions, 144–49, 225;
    "Nobel Gas," 224
tauopathy, 244–45; primary, 245;
    secondary, 245
tau proteins, **234**, 236, 237, 238, 242,
    244–47, 249, **251**, 258, 259
TDP43 prions, 246, 247
Telling, Glenn, 207–8, **208**
Temin, Howard, 269*n*6
Third International Congress of
    Virology (1976), 47
Thomas, Lewis, 85; "The Planning of
    Science," 77
thymine, 24
Time magazine, 195
titin, 99
titration, 26, 65; scrapie, 26–32, 45, 65
titubation, 63
tobacco mosaic virus (TMV), 111,
    112, 113, 121–22
Tokyo, 158, 249
transforming principle, 96
transgenes, 162–60, 176, 207–9,
    235–42, 271*n*10
transgenic (Tg) mice, 162–69, 207–9,
    235–42, 248–49, 250; bioluminescence imaging of, 240–42
transmissible mink encephalopathy
    (TME), 206
transmissible spongiform encephalopathy (TSE), 17, 223–24
Tremblay, Patrick, 216
tremors, 62–63